Marine Mammals in a Changing World

Marine Mammals in a Changing World

Editor

Alexander J. Werth

MDPI • Basel • Beijing • Wuhan • Barcelona • Belgrade • Manchester • Tokyo • Cluj • Tianjin

Editor
Alexander J. Werth
Hampden-Sydney College
USA

Editorial Office
MDPI
St. Alban-Anlage 66
4052 Basel, Switzerland

This is a reprint of articles from the Special Issue published online in the open access journal *Oceans* (ISSN 2673-1924) (available at: https://www.mdpi.com/journal/oceans/special_issues/marine_mammals).

For citation purposes, cite each article independently as indicated on the article page online and as indicated below:

LastName, A.A.; LastName, B.B.; LastName, C.C. Article Title. *Journal Name* **Year**, *Volume Number*, Page Range.

ISBN 978-3-0365-2998-1 (Hbk)
ISBN 978-3-0365-2999-8 (PDF)

Contents

About the Editor

Alexander J. Werth is Trinkle Professor of Biology at Hampden-Sydney College in Virginia, USA. He earned his degrees from Duke University and Harvard University and worked as a postdoctoral researcher with the Department of Wildlife Management in Utqiagvik, Alaska, USA. His research focuses on the functional morphology and biomechanics of feeding in marine mammals (especially suction feeding in toothed whales and filter feeding in baleen whales), but he also publishes research on the evolution of complexity, evolution pedagogy, and the philosophy of science.

Preface to "Marine Mammals in a Changing World"

Marine mammals (e.g., cetaceans (whales, dolphins, and porpoises), pinnipeds (seals, sea lions, and walruses), sirenians (manatees and dugongs), sea otters, and polar bears) are fascinating animals whose bodies, behaviors, and ecology reflect the ways in which their ancestors, long ago, left the terrestrial habitats of typical mammals to revert to the marine habitat. As these marine mammals evolved, and continue to evolve, every aspect of their biology has changed, especially their physical form, anatomy, and biomechanics as well as their physiological and biomechanical processes, including immunological and other crucial responses to environmental stressors. However, their marine environment continues to change in ever more rapid and unexpected ways due to anthropogenic (human) impact. The five published articles in this Oceans Special Issue all involve studies of how marine mammals continue to adapt to their changing environment and how advances in science better enable marine mammal researchers to learn about these changes in marine mammals and their ecosystems. The research article by Cormac Booth and Len Thomas explores the potential adverse health effects of low salinity (from coastal development or other anthropogenic impact) on bottlenose dolphins in addition to the importance of expert knowledge in projecting varied time response scenarios. The research article by Stephanie Norman and colleagues focuses on resistance to antibiotics in harbor porpoises and harbor seals of the Salish Sea in the Pacific Ocean. This resistance, often to multiple drugs, has been caused by environmental contamination from antibiotic and other antimicrobial compounds used in human and veterinary medicine. The research article by Cindy Elliser and colleagues examines many cases of harbor porpoise mortality via asphyxiation from ingestion of large prey fish (American shad). Female porpoises appear to be at greatest risk because they engage in risky foraging behavior in an attempt to maximize caloric ingestion to compensate for energetic needs. The review article by Marie-Anne Blanchet and colleagues looks at changes in harbor seal population status and biogeographic distribution due to multiple emerging threats caused directly or indirectly by human activities. These include climate change, availability of prey species and competition with other predators, shifts in disease-causing pathogens, increases in water contaminants, and increased exposure to pollutants from freshwater runoff. Finally, the review article by Alexander Werth considers the long view of cetacean evolutionary history from the perspective of multiple modern research perspectives, offering hundreds of specific examples of evolutionary topics affecting extant and extinct whales, dolphins, and porpoises while citing detailed research studies outlining investigations for each topic. Together, these articles highlight varied marine mammal species, regions, and research perspectives, but the common thread uniting these articles is change—whether over the sweep of evolutionary time or in short-term accommodation to new threats to the rapidly changing marine habitats where cetaceans, pinnipeds, and other marine mammals are found. Like these marine habitats, marine mammals will continue to change and evolve over time—so long as their populations continue to survive in our ever-changing world.

Alexander J. Werth
Editor

oceans

MDPI

Review

Cetaceans as Exemplars of Evolution and Evolutionary Ecology: A Glossary

Alexander J. Werth

Department of Biology, Hampden-Sydney College, Hampden-Sydney, VA 23943, USA; awerth@hsc.edu

Received: 1 May 2020; Accepted: 21 May 2020; Published: 25 May 2020

Abstract: Extant cetaceans (whales, dolphins, and porpoises) and their extinct ancestors offer some of the strongest and best-known examples of macroevolutionary transition as well as microevolutionary adaptation. Unlike most reviews of cetacean evolution, which are intended to chronicle the timeline of cetacean ancestry, document the current knowledge of cetacean adaptations, or simply validate the brute fact of evolution, this review is instead intended to demonstrate how cetaceans fittingly illustrate hundreds of specific, detailed terms and concepts within evolutionary biology and evolutionary ecology. This review, arrayed in alphabetical glossary format, is not meant to offer an exhaustive listing of case studies or scholarly sources, but aims to show the breadth and depth of cetacean research studies supporting and investigating numerous evolutionary themes.

Keywords: Cetacea; whales; dolphins; porpoise; evolution; macroevolution; paleontology; genetics

1. Introduction

Perhaps no story within the field of evolutionary biology has attracted more popular attention over the past half century as the evolution of whales. This is undoubtedly due to the steady stream of striking, significant, and substantial fossil finds, and to the general appeal of whales and dolphins to scientists and non-scientists alike. No general textbook of modern biology is complete without at least a minor section or box feature outlining the reversion of early cetaceans to the watery habitat of their pre-mammalian tetrapod ancestors, and the many consequent anatomical and ecological changes that followed this major shift [1–5]. This is a story well and broadly told in print, online, and in superb, instructive video documentaries. Where textbooks of the preceding century could reliably be counted on to depict evolution with the history of horses from *Eohippus* to *Equus*, cetaceans are now justifiably cited as prime exemplars of biological evolution. Given this ubiquity, Thewissen and Bajpai [6] crowned cetaceans as the current "poster child for macroevolution".

This paper is intended neither to reiterate the utility of cetaceans in validating the brute fact of evolution—a point well made in numerous popular books [7–10], magazines [11–13], and websites [14–18], and underscored by many excellent museum exhibits and other public resources—nor to review the current knowledge of cetacean ancestry, a timeline chronicled by a growing array of scholarly and popular works [19–33]. Instead, the aim is more narrowly targeted: To show how cetaceans aptly demonstrate specific evolutionary topics. I outline numerous examples of ways in which cetaceans fittingly illustrate detailed terms and concepts within evolutionary biology and evolutionary ecology. These are presented in alphabetical glossary form. They can be used by teachers or scholars searching for examples, or they can simply raise awareness about cetacean research. Neither the examples nor the cited references are meant to offer an exhaustive listing of case studies and illustrations. Rather, the aim is to provide readers, specialists, and non-specialists alike, with an appreciation for the breadth and depth of cetacean research studies.

Given their relatively rapid return to the sea, and hence major change in environment, it is instructive to examine all features of Cetacea within light of the terrestrial-to-aquatic transition.

For example, the multi-chambered cetacean stomach is well known. Is this a legacy of cetacean ancestry (specifically their close relation to—indeed, their classification within—Artiodactyla, many of which have stomachs with multiple compartments), or does this instead represent a functional adaptation: A mill for gastric breakdown of ingested food items in the absence of cusped teeth and mastication typical of mammals?

Like the compartmentalized stomach, many aspects of cetacean bodies and life history offer prime examples for explicating and elucidating evolution. The following list of examples runs the gamut from anatomical, behavioral, genetic, and physiological traits, all intended to demonstrate the ease and effectiveness with which Cetacea provides a deep, rich well of exemplars for teaching and studying evolution.

2. Examples of Evolutionary Terms and Concepts

Adaptation: There are dozens of excellent articles and books in both the classic and contemporary literature explaining numerous examples of cetacean adaptations. Chief among these are explanations of skull telescoping [34], the origins of echolocation and related changes to the ear [35–38], thermoregulation [39], different visual pigments in the eye [40], and different types of myoglobin and other respiratory pigments that bind oxygen [41,42].

Adaptive radiation: Adaptive radiations have been important in cetacean evolutionary history, both in the initial appearance of cetaceans from (presumed) raoellid ancestors as well as in other periods, such as the origin of Odontoceti and Mysticeti (for example, with the evolution of filter feeding) as well as smaller subgroups such as individual genera (e.g., *Stenella* dolphins) or families (e.g., beaked whales, Ziphiidae), or superfamilies (e.g., Delphinoidea) [31,43–46].

Aging/lifespan: The discovery of an old stone point embedded in a bowhead whale, *Balaena mysticetus*, launched a thriving subfield of cetacean studies focused on the remarkably long lifespan (100+ years) of mysticetes, with obvious ramifications for cetacean evolution [47,48]. Such research includes examination of the wax "glove finger" of the mysticete ear, isotopic studies of baleen, or racemization of eyeball amino acids.

Albinism: Albinism has been observed in many cetacean species including sperm, humpback, and killer whales and bottlenose dolphins [49]. This trait is normally governed by a simple genetic mutation, and demonstrates the impact of body coloration and social organization on survival.

Allen's rule: One of several so-called bioclimatic "rules", Allen's rule explains that organisms living at higher latitudes (with cooler climates) tend to have smaller extremities and appendages and thus have less relative surface area over which heat can be lost to the environment. This can be seen with bowhead whale flukes and flippers.

Alloparenting: Alloparenting behavior has been documented in several cetacean species, such as sperm whales that "babysit" juveniles while parents dive deep to forage, or bottlenose dolphin "aunts" (which may or may not be genetically related to true mothers) which help to rear young animals [50–52].

Allopatry: Although it is difficult for some people to see how marine species can be fully isolated geographically (i.e., in disparate and non-contiguous distributions) and thus open to allopatric speciation, demographic studies of whale and dolphin populations support cetacean allopatry [53].

Altruism (intraspecific): Altruistic behavior, which benefits another individual at potential cost to the animal performing the behavior, has been documented in many cetacean species, raising important evolutionary questions about the social circumstances that underlie the possible roots of moral behavior. Dolphins in particular have attracted attention for their apparent altruism to conspecifics, suggesting they may be reciprocal altruists [54].

Altruism (interspecific): Many examples of possible or probable altruistic behavior have also been documented in which a whale or dolphin comes to the aid of individuals of another species (as in epimeletic or care-giving behavior, listed below). This includes many anecdotes from both contemporary and ancient, classical times of whales or dolphins protecting smaller, vulnerable animals, such as seals or human swimmers, from drowning or from predators such as sharks [55]. The extent to which such behaviors might be instinctive, and to which cetaceans performing such behaviors recognize that the animals they are aiding are not members of their own species, is the object of much speculation.

Anagenesis: The possibility of cetacean taxa evolving via connected "straight-line" evolution (i.e., without cladogenesis) has been discussed [56].

Analogy/analogous features: Apart from examples of convergent evolution within Cetacea (see below), there are many obvious examples of cetaceans sharing analogous features (i.e., bearing similar form or function yet without a common evolutionary basis) with other taxa. Classic examples include the general fusiform body shape and stabilizing dorsal and pectoral plus propulsive caudal fins in sharks, ichthyosaurs, and dolphins [1–3].

Ancestral state reconstruction: Given their striking evolutionary history and increasingly well-known fossil record, reconstruction of ancestral states of cetaceans have been involved in many projects, often focused on body size (and in particular gigantism; see below) [57] but also involving changes in feeding mode over time [58,59].

Apomorphy: Examples of novel, derived traits that distinguish taxa, such as Cetacea or Mysticeti, include many detailed studies of ear bones [35–38,60,61], as well as analyses of dental and baleen morphology [62], plus highly derived bones or other cranial features related to feeding.

Atavism: The "reappearance" of "lost" hindlimbs or rudimentary yet enlarged pelvic elements are among the most notable and distinct examples of atavisms in any living animals [63].

Bergmann's rule: Like Allen's rule (see above), this bioclimatic "rule" holds that body size and shape vary by latitude, in this case with larger animals being found at higher latitudes, such that they have a relatively lower surface-to-volume ratio and correspondingly less heat loss from the body to cold ambient waters. The stout bowhead whale—which is not only more rotund than whales of other families but also of closely related right whales—is a prime example of this finding.

Biogeographical distribution: There are several good examples of cetacean distribution relating to evolution, such as the antitropical distribution of corresponding northern and southern species (such as right whale dolphins) [64], as well as the riverine and estuarine distribution of closely related or convergent taxa (such as river dolphins) [65].

Biological magnification: Cetaceans offer prime examples of biomagnification due to accumulation in tissues of neurotoxins from "red tides" as well as many types of anthropogenic pollutants including methylmercury and organochlorines [66–69].

Biostratigraphy: As is typical throughout paleontology, the fossil layers (geologic strata) in which cetacean fossils are found offer numerous clues to aid in dating the fossils as well as establish paleoecological or climatological information, making biostratigraphy a common and essential element of any fossil study [70].

Bone bed formation: Although some whale fossils, including whole skeletons, may be found isolated from bones of other individuals, the bones are frequently found in mixed "bone beds" containing fossils of multiple individuals, and sometimes multiple species. This is true of some of the oldest known archaeocetes as well as more recent whales and dolphins in many rich fossil localities, such as the Sharktooth Hill bonebed from the middle Miocene of California [71].

Capital breeding: As opposed to income breeders that use exogenous energy sources to "finance" reproduction, capital breeders build up and maintain internal energy stores before reproducing, a state that has been characterized for many mysticete species [72,73].

Carbon pump: The "fertilization" of primary production in epipelagic seas by the return of carbon, nitrogen, and other nutrients to surface waters via whale feces, has been documented and much discussed in recent years [74,75], with major implications not only for global nutrient cycles but also the key role of pre- and post-whaling cetacean populations in modulating and regulating oceanic ecosystems.

Character displacement: Basic studies of resource partitioning in cetacean taxa include studies of body size, skull telescoping and migration of external nares through evolution, and head shape and dental loss in many odontocete species [76,77].

Chronospecies: The idea that a single species might evolve directly (without divergence) into a morphologically distinct form, yielding a single species (or closely related sister species) connected over time in the fossil record (see entry for anagenesis), has been discussed for mysticetes such as minke whales [78].

Circumpolar: The bowhead whale is a good example of a single cetacean species with multiple breeding stocks in circumpolar distribution (e.g., from the Arctic Ocean to North Atlantic or North Pacific), which has clear implications for speciation, as the now-extinct population of Atlantic gray whales attests [79].

Coevolution: The contemporary and linked evolution of cetaceans with ectoparasitic whale "lice" provide a clear and strong example of host–parasite coevolution [80].

Coloration patterns: There have been several analyses of pigmentation and its relation to character displacement and species recognition, as well as to confusing or startling prey (as by the bright white flippers of humpback whales or asymmetrically colored jaws of fin whales); see countershading entry below [81].

Competition: Competition plays a major role in ecology and evolution, whether intraspecific, such as sperm competition between conspecific males [82], or interspecific, such as between whales and penguins and other species for krill and other food [83].

Constraints: Evolutionary outcomes depend on raw materials but also intrinsic (genetic) and extrinsic (environmental) constraints, which may explain why suspension feeding evolved in cetaceans but not marine reptiles [84], although a plesiosaur has recently been described that possibly was a filter feeder [85].

Convergence: There are many examples of convergent evolution within Cetacea but perhaps none more striking than the tusked, walrus-convergent *Odobenocetops* of the Miocene and Pliocene, whose skull, dentition, and presumed lifestyle appear to have closely mirrored that of the living walrus, *Odobenus* [86,87]. Another fine example of convergence involves a beaked whale-like Pliocene dolphin excavated from Antarctica [88].

Cooperation: Many cooperative behaviors, ranging from cooperative foraging to defense against predators, have been documented in cetaceans [89]. Some of these involve complex and instinctive or possibly learned traits. There also appears to be cooperative fishing with humans by some dolphins.

Countershading: A common coloration pattern in aquatic as well as terrestrial and aerial species, especially large predators, leads to light coloration on the animal's underside (so that it blends in with light from above) and dark coloration on the dorsal surface (so that an animal is hard to detect when viewed from above), and this countershading is prevalent in many cetacean species [81,90].

Culture: Several aspects of distinct cultural transmission of behaviors have been described in odontocete and mysticete species [91,92].

Death and dying behavior: There have been numerous documented instances of dolphins or other cetaceans carrying bodies of dead offspring, or of close attention paid to dead animals by various related and unrelated individuals in many cetacean species [93].

Degeneracy/"devolution": The loss of complex enamel structure (typical of nearly all non-cetacean mammals) is likely linked to the loss of dental function (typical mammalian mastication) in odontocetes; this loss of complexity appears to be an example of "devolution" [94,95]. Because such losses represent potential decline of reversal of evolution, they are often referred to as degeneracy.

Demography: The possible role of post-menopausal females in cetacean populations has spurred demographic studies focused on population breakdowns and genetics [96].

Disjunction distribution: The majority of whales, dolphins, and porpoises have disjointed (noncontinuous) geographic distributions, also known as range fragmentation. This is significant for demographic, genetic, and evolutionary reasons as well as ecological conditions [97,98].

Disruptive selection: An example of disruptive natural selection in cetaceans, apart from the obvious cases of ecological divergence based on foraging or biogeographic distribution, may involve a natural "knockout" of a basic cytokine protein that acts as an immune signal [99].

Dwarfism: The pygmy blue whale and dwarf and pygmy sperm whales are examples of subspecies and related species, respectively, that demonstrate evolutionary changes in body size (see also gigantism).

Ecotypes: One of the best examples of evolutionary divergence and resource-based character displacement involves the diversity of ecotypes among killer whales, long thought to represent a single species (*Orcinus orca*) but which may represent distinct species or subspecies in addition to clear ecotypes, such as stocks that feed mainly on salmon or other bony fish, those that prey largely on large sharks, and those that feed mainly on other marine mammals, including small porpoises and seals as well as huge whales [100,101].

Embryology: Although some morphologists studied cetacean embryos and fetuses hundreds of years ago, there is now a much more systematic study of the development of cetacean species, and particularly by using modern molecular and histological methods [102,103].

Encephalization and brain evolution: The story of cetacean brain evolution has attracted much attention in recent years, especially with regard to comparisons with the evolution of large, complex brains in hominins and other primates [104]. This promises to be fertile ground for evolutionary studies.

Endangered/vulnerable species: Several cetacean species have, for various reasons (largely whaling, ship strikes, fishing gear entanglement, and habitat destruction), been and in some cases continue to be highly endangered, including the North Atlantic right whale and the vaquita (a porpoise endemic to the Sea of Cortez which is nearly extinct).

Endemic: Like many organisms, cetaceans demonstrate endemicity, being confined to particular regions. In Cetacea this most commonly occurs in riverine or coastal dolphins [105].

Epigenetics: Apart from various means used to age individual whales and dolphins, such as accumulation of layers of dental tissues or clock-based degradation of molecules [106], epigenetic explanations (i.e., beyond genes) have been proposed to explain how cetaceans may fight cancer [107].

Epimeletic behavior: Care-giving behavior may include the altruistic behaviors described above (such as saving people or other animals from drowning or protecting them from predators) or simply devoting much care and attention to unrelated individuals [108,109].

Evolution of complexity: Just as the loss of odontocete enamel relates to loss of complex structure and function, the evolution of echolocation and vocalizing structures and behaviors in various cetaceans relates to the gain of complexity [33,36,95,110,111], as does the origin and evolution of baleen.

Evolutionarily stable strategy (ESS): Unlike an evolutionarily stable state, an ESS is a behavioral strategy that is fixed or cannot be "invaded" by an alternative gene-based behavior. Various altruistic and other cultural behaviors of dolphins or other cetaceans may qualify [54].

Evolutionary turnover: According to the turnover-pulse hypothesis, major environmental changes often spur major turnover and adaptive radiation of taxa, as is presumed to have occurred during the spread of Neoceti (crown cetaceans) due to Oligocene oceanographic changes [44,112].

Evolutionary-developmental biology ("evo-devo"): The role of regulatory genes leading to morphological changes in dentition and hindlimbs has been the subject of several studies [30,113,114].

Evolvability: The capacity of cetaceans for adaptive evolution due to their molecular and morphogenetic changes after leaving behind their terrestrial ancestry has attracted attention, often involving osteological changes related to ears and hearing or other parts of the cranium [115].

Exaptation: Much speculation about exaptation ("preadaptation") in cetaceans focuses on the hearing capabilities of the first cetaceans, which were amphibious and had water-adapted hearing that "exapted" them to evolve further into more fully aquatic habitats, and which led to diverse vocalizations (with corresponding brain and anatomical specialization) that ultimately led to echolocatory abilities [116,117].

Extinction: There have, of course, been many cetacean taxa that went extinct and are now known solely from fossil material, but there has also been a recent instance of a living species (the baiji or Yangtze River dolphin, *Lipotes vexillifer*) which was recently declared extinct [118].

Finite element analysis (FEA): Bite strength in living and extinct cetacean taxa has been analyzed by FEA [119].

Fission–fusion structure: Although more commonly studied in primates, this social structure (in which individuals of a species temporarily join, then go separate ways) has also been documented in various cetacean taxa [120].

Fitness: Among the many examples of evolutionary fitness in cetaceans, the role of the narwhal tusk stands out as an interesting and good example [121,122].

Food fall: There have been many studies investigating the trophic and other ecological roles of cetacean carcasses that decompose over many years on the seafloor (see entry on whalefall communities). These involve observations, experiments, and even analyses of fossil material [123,124].

Fossil dating: Fossils have been used to date divergence ages of extant lineages [125], and multiple methods have been employed to determine the age of cetacean fossils, ranging from traditional isotopic analysis and other molecular methods to geologic formation analysis and use of index fossils [126,127].

Fossil lagerstätte: A lagerstätte is a fossil locality with high diversity, often with numerous complete skeletons (such as the "valley of whales" bearing many basilosaurids and protocetids in Fayum, Egypt), and excellent quality of preservation, such as the Pisco Basin of Peru, where conditions led to preservation of baleen and even digestive tract contents [128,129].

Fossil reconstruction: Many aspects of cetacean form have been reconstructed in extinct taxa, ranging from body size and shape to the curvature and proportions of the spine and their role in locomotion [56,57,130,131].

Genetic drift—bottleneck: The effects of severe population size decrease, namely seemingly random fluctuations in allele and genotype frequencies, have been studied in populations of mysticetes and odontocetes [132,133].

Genetic drift—founder effect: Similar to bottlenecks, small groups can also have random genetic effects when new populations are founded by a very few individuals; this has been studied in various dolphin species [134,135].

Genomics: Several studies have looked at the evolution of the whole genome in particular cetacean species plus higher-level taxa [136–138].

Gigantism: Multiple recent studies have looked at the evolution of extreme body size in Cetacea with relation to various factors such as trophic ecology or biomechanics and morphology [139–141]. This includes not only the obvious mysticetes but also gigantic toothed whales [142].

Group selection: The concept of group selection (and levels of selection in general) is controversial in evolutionary biology, but studies of whale and dolphin sociality relate directly or indirectly to this topic [92,143].

Habitat loss: Sadly, there are many examples of the evolutionary effects of habitat loss affecting diverse cetaceans in prehistoric and modern times—with the latter obviously involving human impact [144].

Homology: Among the many obvious examples of homologous morphological structures in cetaceans are varied bones [145] and teeth [146]; there are of course homologous chromosomes and genes too.

Host–parasite interactions: There are many records of endo- and ectoparasites on and within different cetaceans, but also interesting evolutionary stories of "switching" of hosts inferred by DNA [147]. See entries below on parasitism and phoresis.

Human impact: Humans have played and continue to play a large role in influencing the evolution of cetaceans through such acts as whaling, driving climate change, and destroying habitats (see habitat loss) [148].

Human impact-anthropophily: A fascinating story of dolphins adapting to, and working cooperatively with, human fishing (in multiple locations around the world) reflects the roles of genes, instincts, and learning in driving cetacean social and behavioral evolution [149,150].

Hybridization: There have been widely reported instances of interspecific hybrid "wholphins" in captivity, but also numerous documented cases of hybrid dolphins and large whales (e.g., blue/fin whales) in natural habitats, spurring speculation as to speciation and genetic divergence [151].

Infectious disease evolution: Changes over time in frequency or at least documentation of various diseases in wild cetacean populations has focused not only on the diseases but also the role of human impact (from pollution and poor sanitation, etc.) in altering the evolution of these diseases and the way they infect whales and dolphins [152].

Irreducible complexity: Critics of evolution often argue that many traits are too complex to have evolved. Traits such as the large brains and echolocatory abilities have been mentioned among cetaceans. Darwin himself speculated, in the first edition of the *Origin of Species*, about the evolution of baleen and complex filtering form, function, and behavior from swimming bears catching aquatic insects that he presumed might someday evolve into whale-like creatures [153].

Iteroparity: Unlike semelparous organisms that reproduce just once before dying, iteroparous organisms have multiple reproductive events over their lifespan; in cetaceans the timing of this often depends on energy state and accumulation of nutrients [73].

Key innovation: There are many obvious examples of key innovations that ushered in major changes in the ecology and evolution of cetaceans, including the origins of baleen, echolocation, large brains, and structures associated with producing and receiving sound waves (such as the melon, mandibular "pan bone," and inner ear). The axial skeleton has also been offered as an example of a key innovation important to cetacean evolution [154].

Keystone predators: Killer whales have been proposed as a classic example of a keystone predator whose action helps to regulate the population dynamics and ecology of multiple species (from fish and sea otters to seals, sea lions, and other odontocetes) in marine ecosystems [155].

Kin selection: Alloparenting and related caregiving behaviors by sperm whales and post-menopausal "grandmother" killer whales has been posited as potentially being related to kin selection [96,156].

K strategy: Whales, dolphins, and porpoises are often presented as classic examples of the K-selected life history strategy favoring large bodies with slow growth and maturity, long lifespan, great devotion of resources to few offspring, and so on.

Life history: Analysis of the ways in which various life history factors (such as body size, lifespan, age of maturity, number of offspring, and so on) relate to cetacean evolution has been conducted [157].

Living fossil: The pygmy right whale, *Caperea*, has been proposed as a remnant of an otherwise ancient and extinct family of early mysticetes, the cetotheriids [158], although it appears that cetotheres persisted into the Pleistocene [159]. The Ganges river dolphin, *Platanista*, is similarly a remnant of a formerly diversified clade (Platanistoidea).

"Lumpers vs. splitters": These colloquial terms refer to the preferences among systematists to classify taxa into as few or as many species (or other taxonomic ranks) as possible. Depending on one's view, there may be 75–120 different extant cetacean species, with much of the disagreement involving dolphins, beaked whales, and rorquals of the genus *Balaenoptera*.

Mating and social systems: The intricacies of mating systems among diverse whale and dolphin species are often complex (befitting their social complexity) and interesting, as with the intense sperm competition of right whales, and have attracted much scrutiny [160,161].

Metonym (taxonomic synonym): There have been several instances of systematists taking names of extant or extinct cetacean taxa and reusing them to apply to a new taxon [162].

Migration: The relation of long migrations undertaken by whales and dolphins for mating, feeding, and other important activities related to survival and reproduction has been studied not only for specific taxa but in general terms [163].

Mimicry: A frequently cited instance of likely mimicry in Cetacea involves the shark-like appearance (with underslung jaws bearing sharp teeth, plus pigmented false gill slits) of dwarf and pygmy sperm whales, *Kogia*, although there are also many mentions of possible vocal mimicry.

Modularity: The concept of modularity, in which a structural or functional system can be subdivided into sets of autonomous yet interacting elements (which are altered and interrelated via "evo-devo" changes), has received much recent attention within evolutionary biology. This includes examples within Cetacea [164].

Molecular clock: Many estimates of the origins, divergences, and lifespans of various cetacean taxa have been derived from molecular data by many researchers [136,165–167].

Morphological disparity/phenotypic diversity: Many cetaceans display remarkable disparity (e.g., members of the beaked whale genus *Mesoplodon*), which has been used to study the evolution of Cetacea [44].

Morphological vs. molecular data: The long-standing issue of agreement between anatomical (often osteological) and molecular findings in resolving phylogenetic issues also includes several thorough analyses of cetaceans [117,140,168,169].

Morphometrics: Detailed morphometric studies of nearly every conceivable aspect of cetacean form have been carried out, ranging from overall body size and shape (as relates to locomotion or thermoregulation) as well as teeth, skin, brains, and varied skull structures; several such studies relate directly to phylogenetics and evolution [170].

Mosaic evolution: The extent to which cetacean form and function represents a blend of ancestral and derived characters has been considered in multiple studies involving various organs such as the brain [171]. The gradual loss of hindlimbs in archaeocetes and transition of the forelimb into the flipper also involve mosaic evolution.

Mutation: Along with general exploration of mutations involved in the terrestrial-to-aquatic transition of early cetaceans [1–3], many specific studies have examined specific gene mutations and their consequences in cetaceans, most notably involving key events in cetacean evolution (such the loss of body hair) and mutations related to olfaction, gustation, vision, and other senses [172–174].

Mutualism: Among the many described instances of mutualism in Cetacea are cases involving whales and non-cetacean taxa (such as seabirds, where the interaction may involve cleaning of parasites from whales as well as location of food sources), as well as discussions of mutualistic interactions (for example, for feeding or defense against predators) involving different cetacean species including interactions between dolphins and large whales [175,176].

Natural vs. artificial selection: The extent to which large-scale twentieth century industrial whaling may have inadvertently altered whale behavior, size, ecology, distribution, and so on, affords an excellent opportunity to compare the effects of human versus natural influences on evolution.

Neoteny and pedomorphosis: Among the many recent investigations focused on changes in developmental timing (see "evo-devo"), and in particular the retention of juvenile features, are comparative studies that closely examined the skulls of extant and extinct whales, dolphins, and porpoises [177,178].

Neuroscience: Outside of Primates, Cetacea is one of the most actively studied groups within the burgeoning field of evolutionary neuroscience, with many projects and publications looking at absolute and relative brain size, the organization of neural networks and brain regions, and the relationship between brain and behavior, including vocalization, sensation, and sociality.

Neutral theory: Many mutations within Cetacea are presumed to have had little to no effect on fitness, yet may elucidate phylogenetics or demographics [179].

Niche separation: The partitioning of food and other resources by contemporaneous humpback and minke whales in Antarctic waters offers a prime example of niche separation and competitive exclusion in Cetacea [180].

Nomenclature: The Latin binomials of many cetaceans—such as the beluga (*Delphinapterus leucas*, or "white dolphin without a fin"), narwhal (*Monodon monoceros*, or "one tooth, one tusk"), and humpback whale (*Megaptera novaeangliae*, or "big-winged New Englander")—offer good lessons in the principles and practice of naming taxa for experts and beginners alike, as do the common names of these species, along with others (killer whale vs. orca, rorqual, etc.).

Nutrient distribution/trophic connections: In addition to several recent studies that have looked at the role of whales in distributing nutrients throughout marine ecosystems, other investigations have explored trophic interactions between cetaceans and other marine predators (e.g., sharks and penguins) for food [83].

Opportunism: Whereas some cetaceans appear to be highly specialized, others, such as the bottlenose dolphin, *Tursiops* spp., are successful ecological opportunists, with obvious evolutionary ramifications.

Organ systems: All cetaceans offer prime examples of organs and organ systems highly modified by evolution, such as kidneys that adapted to the switch from a terrestrial to marine environment (with corresponding lack of fresh drinking water), or the lungs and diaphragm modified for greater tidal volume and more efficient pulmonary ventilation, etc. Vascular (often retial) adaptations for diving and thermoregulation are also excellent examples of fundamental evolutionary changes.

Orthogenesis: Studies of cetacean evolution and diversity have not provided evidence for the claim of progressive, directed (i.e., non-random) evolution, although the concept has been discussed [181].

Osteological correlates: The study of bony landmarks and their significance in denoting major functional changes important during cetacean evolution (such as palatal sulci relating to vasculature for baleen, or muscle scars relating to origin/insertion attachment points) has proven invaluable in cetacean paleontology and morphology.

Outgroup comparison: Numerous studies have affirmed the relationship of Cetacea within Artiodactyla (or Cetartiodactyla), with hippopotamuses as the outgroup [182]; other studies have examined outgroups within Cetacea, such as the placement of porpoises within Delphinoidea.

Pair bonding: The tucuxi (*Sotalia* spp.) is sometimes offered as an example of a cetacean with a pair-bonded mating system [183].

Paleoecology: Among the many studies of cetacean paleoecology are fascinating stories of the likely predation by the extinct giant shark *Carcharocles megalodon* on baleen whales of all sizes [184].

Parallel evolution: Comparisons between bats and odontocetes as a good example of parallel evolution of echolocation are common [185].

Paraphyly: Paraphyletic groups have been noted in cetacean systematics, especially with older classifications of river dolphins, and more recently with genetic analyses of delphinine dolphins including *Stenella* and *Tursiops* [186].

Parasitism: Cetaceans are definitive hosts for numerous ecto- and endoparasites, including *Anisakis* worms, which easily spread to humans who eat raw or undercooked fish.

Phenetic vs. cladistic systematics: Just as there have been comparisons of molecular and morphological findings in resolving systematic and phylogenetic debates about cetacean taxonomy, so too differing results of phenetics (systematics based on similarity of form) and cladistics have yielded different conclusions, and debate [170].

Phoresis: Cetaceans are well known to "carry" (in a sort of commensal mutualism) many species of ectoparasitic barnacles, worms, and whale lice, along with remoras [187]. This is related to parasitism (see entry above).

Play behavior: Play, considered an important element and indicator of complex social interaction and cognitive ability, has been documented within numerous wild and captive cetaceans [188].

Pleiotropy: The reproductive tracts of some cetaceans may demonstrate pleiotropic genetic interactions [189].

Plesiomorphy: Enamel patterns [190] within cetaceans have been cited as an example of ancestral (plesiomorphic) traits persisting in modern taxa; to a lesser extent, ear bones, although highly modified in Cetacea, also demonstrate some plesiomorphic features [60,61].

Polydactyly vs. polyphalangy: Whales largely retain the plesiomorphic condition of five digits in the forelimb (flipper), although there are instances of digit reduction in cetaceans [191]. Cetaceans, like many other aquatic tetrapods, also are unusual in having an abnormally large number of phalangeal bones within each digit, such that flipper osteology offers a good example of mosaic evolution.

Polygyny: Polygynous mating systems, in which one male has access to a "harem" of several reproductively receptive females, can be found in many cetacean species ranging from sperm to humpback whales [192].

Polymorphism: There are many examples on (and many published papers on) diverse morphological, behavioral, and molecular traits within Cetacea [193].

Polyphyly: The diverse dolphin genus *Lagenorhynchus* is sometimes offered as an example of a potentially polyphyletic taxonomic grouping within Cetacea [132,193].

Rate of evolution: The extent to which evolution occurs via incremental gradualism or sudden leaps plays a role in studies of neutral mutations, molecular clocks, and regulatory gene interactions (see entries for all items in this list), and cetaceans have provided fodder for interpretations of both slow and rapid evolutionary change, often dependent on environmental factors and the origin of key innovations [194].

Red Queen hypothesis: A spiraling evolutionary "arms race" between cetacean immune systems and pathogens, possibly including neurotoxins from paralytic shellfish poisoning, has been cited as an example of Van Valen's famous "Red Queen hypothesis," in which taxa must "keep moving just to stay in one place" [56,195].

Regulatory genes: Homeobox and other regulatory genes (like *sonic hedgehog* and *Hox*) have been cited as important in major evolutionary transformations throughout the history of Cetacea involving form and function, such as limb and dental loss and the origins of baleen and echolocation [137].

Relict populations: Diverse river dolphin taxa as well as some oceanic dolphins, notably the rough-toothed dolphin (*Steno*), have been proposed as ancestral relict populations throughout their distribution or in some locales [196].

Reproductive isolation: Reproductive isolation between cetacean populations and subpopulations have been cited as important steps toward speciation or other genetic and cultural divergence [60,64,197].

Reproductive senescence: Like some higher primates, several cetacean species have been noted as having frequent and important post-reproductive phases, including "grandmother" killer whales that potentially pass along cultural knowledge [198,199].

Resource partitioning: There have been several published examples of resource partitioning within Cetacea (e.g., of different Antarctic whale species partitioning prey or feeding habitat) as well as of cetaceans partitioning resources with other marine species including sharks and seabirds [200].

Ring species: Although there is no definitive example of a cetacean ring species, the presence of numerous subspecies, interspecific hybrids, and intergraded populations of widely dispersed species (often with circumpolar distributions, as in killer whales) makes the prospect of the ring species concept within Cetacea a distinct possibility.

Scaling (isometry vs. allometry): Dozens of publications demonstrate the roles of linear and nonlinear scaling effects in the evolution of cetacean structures ranging from bones, limbs, organs, tissues (such as skin thickness), and other features. These indicate that scaling has played a prominent role in cetacean polymorphism and phenotypic disparity [201].

Selective sweep: Studies of genetic diversity within various cetacean taxa, including some (like sperm whales) with less diversity than expected, demonstrate the likely role that strong selective sweeps play in fixing alleles within a population [202].

Sensory biology: Although studies of cetacean ears and hearing have long attracted research interest, many recent studies have demonstrated that other sensory modalities (vision, olfaction, etc.) are far more complicated and important in cetacean ecomorphology and evolution than generally appreciated [172–174].

Sexual conflict: The likelihood of diverging male and female reproductive strategies (and counterstrategies) in mating behavior and anatomy has led to a greater recognition of and research interest in applications of sexual conflict theory within Cetacea [203].

Sexual dimorphism: Sexual dimorphism is common in many cetaceans. In mysticetes, females are generally larger than males, but in odontocetes males are typically larger, often dramatically so (e.g., sperm and killer whales).

Sexual selection: The large tusk of narwhals and prominent mandibular teeth of ziphiids (which in many taxa erupt solely in males) are commonly cited examples of sexual selection in Cetacea. These teeth are often considered to play a role in competition for females, either via display or direct male-to-male fighting [56,204].

Speciation: Antitropical distributions have been cited as a means of enabling allopatric speciation in Cetacea, although other rapid genetic changes or morphological and behavioral disparity (leading to reproductive isolation) might in turn lead to sympatric or parapatric speciation [56,64].

Stabilizing selection: Several genes have been described as having been "purified" or fixed and stabilized at high levels within cetacean populations [42].

Stable isotope analysis: For several decades stable isotopes isolated from various tissues (bones, baleen, etc.) have been widely used to indicate many parameters including distinct populations, ecological states (e.g., trophic levels), and physiological condition (e.g., stress, reproductive condition, age, etc.) [205,206].

Subspecies: Genetically and morphologically distinct subspecies have been described for many species of dolphins and whales, including, for example, humpback and fin whales [207].

Symbiosis: Among the many symbiotic interactions involving cetaceans, familiar examples include multiple species of barnacles that depend exclusively on whales or dolphins for dispersal.

Sympatry: Many cetacean species have been described as sympatric, including river dolphins sharing isolated habitats as well as oceanic whales and dolphins with neritic or pelagic habitats [208,209].

Taphonomy: The condition and potential taphonomic alteration of many fossil cetacean materials has been well described [71,210–212].

Tool use: Researchers have described the use of natural sponges or similar materials by dolphins searching for prey in benthic sediments, a behavior that appears to be culturally transmitted [213].

Top-down vs. bottom-up trophic cascades: Researchers have described examples of differing trophic cascades involving whales, dolphins, and porpoises, where either the cetacean (as a large predator) heavily influences the presence and abundance of primary producers and consumers, in a so-called top-down cascade [214], as well as cases of bottom-up cascades where species lower in a trophic pyramid influence the abundance of cetaceans and other large predators [83], such that both types of trophic cascades apply to cetaceans [215,216].

Transitional fossils: Considering the remarkable evolutionary transitions that have occurred throughout cetacean history, such as wholly living, locomoting, and hearing in water instead of air, or capturing prey by filtering with baleen instead of grasping with teeth, there have been many taxa described as classic transitional fossils, indicating forms intermediate between other known forms [1–12].

Vestigial features: Cetaceans are among the best-known and frequently cited examples of organisms displaying vestigial features. These include pelvic and limb bones, and hairs and hair follicles.

Vicariance: Extant river dolphins and extinct cetotheres are among the notable taxa whose vicariance (geographic separation into discontinuously distributed groups) has been studied [217,218].

Whalefall community: As previously noted in the entry on food falls, cetacean carcasses often remain on the seafloor as an important contributor to benthic trophic webs. These dead bodies, which may take years to be fully digested and decomposed, form the basis for distinctly unique benthic "whalefall" communities [123,124].

3. Conclusions

Every group of organisms can be used to highlight specific facets of evolution and evolutionary ecology, but extant and extinct whales, dolphins, and porpoises perhaps demonstrate the breadth of evolutionary topics better than any other taxon. The items listed in this glossary are intended to demonstrate the wide range of topics studied by cetacean scientists (they are not meant to provide a complete, exhaustive listing), and this list will undoubtedly grow as new methods yield new insights and discoveries. As surely as cetaceans continue to evolve, so too the fields of study involving them continue to evolve as well.

Funding: This research received no external funding.

Acknowledgments: I thank my many research colleagues and students who have helped to teach and learn with me. I am especially grateful to Olivier Lambert and two anonymous reviewers for providing helpful suggestions including relevant topics and references.

Conflicts of Interest: The author declares no conflict of interest.

References

1. Kellogg, R. The history of whales: Their adaptation to life in the water. *Q. Rev. Biol.* **1928**, *3*, 29–76. [CrossRef]
2. Howell, A.B. *Aquatic Mammals: Their Adaptations to Life in the Water*; Charles C. Thomas: Springfield, IL, USA, 1930.
3. Thewissen, J.G.M.; Cooper, L.N.; George, J.C.; Bajpai, S. From land to water: The origin of whales, dolphins, and porpoises. *Evol. Educ. Outreach* **2009**, *2*, 272–288. [CrossRef]
4. Reynolds, J.E.; Rommel, S.A. *Biology of Marine Mammals*; Smithsonian: Washington, DC, USA, 1999.
5. Reidenberg, J. Anatomical adaptations of aquatic mammals. *Anat. Rec.* **2007**, *290*, 507–513. [CrossRef] [PubMed]
6. Thewissen, J.G.M.; Bajpai, S. Whale origins as a poster child for macroevolution. *BioScience* **2001**, *51*, 1037–1049. [CrossRef]
7. Zimmer, C. *At the Water's Edge: Fish with Fingers, Whales with Legs, and How Life Came Ashore But then Went Back Again*; Atria Books: New York, NY, USA, 1999.
8. Berta, A. *Return to the Sea: The Life and Evolutionary Times of Marine Mammals*; University of California Press: Berkeley, CA, USA, 2012.
9. Thewissen, J.G.M. *The Walking Whales: From Land to Water in Eight Million Years*; University of California Press: Berkeley, CA, USA, 2014.
10. Pyenson, N.D. *Spying on Whales: The Past, Present, and Future of Earth's Most Awesome Creatures*; Viking: New York, NY, USA, 2018.
11. Mueller, T. Valley of the whales. *Natl. Geogr.* **2010**, *218*, 118–137.
12. Giggs, R. Giants of the Deep. *The Atlantic* **2018**, *9*. Available online: https://www.theatlantic.com/magazine/archive/2018/09/whale-evolution/565760/ (accessed on 1 April 2020).
13. Black, R. How Did Whales Evolve? Smithsonian Magazine. Available online: https://www.smithsonianmag.com/science-nature/how-did-whales-evolve-73276956/ (accessed on 1 April 2020).
14. When Whales Walked on Four Legs. Natural History Museum of London. Available online: https://www.nhm.ac.uk/discover/when-whales-walked-on-four-legs.html (accessed on 1 April 2020).
15. The Evolution of Whales. University of California Museum of Paleontology. Available online: https://evolution.berkeley.edu/evolibrary/article/evograms_03 (accessed on 1 April 2020).
16. Whale Evolution. PBS WGBH Evolution Series Library. Available online: https://www.pbs.org/wgbh/evolution/library/03/4/l_034_05.html (accessed on 1 April 2020).
17. Evolution of Whales Animation. Smithsonian Institution. Available online: https://ocean.si.edu/through-time/ancient-seas/evolution-whales-animation (accessed on 1 April 2020).
18. Whale Evolution from Walking Whales to *Janjucetus*. Melbourne Museum-Museums Victoria. Available online: https://museumsvictoria.com.au/website/melbournemuseum/discoverycentre/600-million-years/videos/whale-evolution/index.html (accessed on 1 April 2020).

19. Thewissen, J.G.M. *The Emergence of Whales: Evolutionary Patterns in the Origins of Cetacea*; Springer: New York, NY, USA, 1998.
20. Mchedlidze, G.A. *General Features of the Paleobiological Evolution of Cetacea*; Oxonian: New Delhi, India, 1984.
21. Marx, F.; Lambert, O.; Uhen, M.D. *Cetacean Paleobiology*; Wiley-Blackwell: London, UK, 2016.
22. Berta, A.; Sumich, J.L.; Kovacs, K. *Marine Mammals: Evolutionary Biology*, 3rd ed.; Elsevier/Academic Press: San Diego, CA, USA, 2015.
23. McGowen, M.R.; Gatesy, J.; Wildman, D.E. Molecular evolution tracks macroevolutionary transitions in Cetacea. *Trends Ecol. Evol.* **2014**, *29*, 336–346. [CrossRef]
24. Gatesy, J.; O'Leary, M.A. Deciphering whale origins with molecules and fossils. *Trends Ecol. Evol.* **2001**, *16*, 562–571. [CrossRef]
25. Gould, S.J. Hooking leviathan by its past. *Nat. Hist.* **1995**, *94*, 8–15.
26. O'Leary, M.A.; Uhen, M.D. The time and origin of whales and the role of behavioral changes in the terrestrial-aquatic transition. *Paleobiology* **1999**, *25*, 534–556. [CrossRef]
27. Gatesy, J.; Geiser, J.H.; Chang, J.; Buell, C.; Berta, A.; Meredith, R.W.; Spring, M.S.; McGowen, M.R. A phylogenetic blueprint for a modern whale. *Mol. Phylogen. Evol.* **2013**, *66*, 479–506. [CrossRef] [PubMed]
28. Deméré, T.A.; McGowen, M.R.; Berta, A.; Gatesy, J. Morphological and molecular evidence for a stepwise evolutionary transition from teeth to baleen in mysticete whales. *Syst. Biol.* **2008**, *57*, 15–37. [CrossRef] [PubMed]
29. Thewissen, J.G.M.; Hussain, S.T.; Arif, M. Fossil evidence for the origin of aquatic locomotion in archaeocete whales. *Science* **1994**, *263*, 210–212. [CrossRef] [PubMed]
30. Thewissen, J.G.M.; Cohn, M.J.; Stevens, L.S.; Bajpai, S.; Heyning, J.; Horton, W.E., Jr. Developmental basis for hind-limb loss in dolphins and origin of the cetacean bodyplan. *Proc. Natl. Acad. Sci. USA* **2006**, *103*, 8414–8418. [CrossRef] [PubMed]
31. Thewissen, J.G.M.; Williams, E.N. The early radiation of Cetacea (Mammalia): Evolutionary pattern and developmental correlations. *Ann. Rev. Ecol. Syst.* **2002**, *33*, 73–90. [CrossRef]
32. Thewissen, J.G.M.; Cooper, L.N.; Clementz, M.T.; Bajpai, S.; Tiwari, B.N. Whales originated from aquatic artiodactyls in the Eocene epoch of India. *Nature* **2007**, *450*, 1190–1191. [CrossRef]
33. Gingerich, P.D.; Wells, N.A.; Russell, D.E.; Ibrahim Shah, S.M. Origin of whales in epicontinental remnant seas: New evidence from the Early Eocene of Pakistan. *Science* **1983**, *220*, 403–406. [CrossRef]
34. Roston, R.; Roth, V.L. Cetacean skull telescoping brings evolution of cranial sutures into focus: Telescoping and cranial suture evolution. *Anat. Rec.* **2019**, *302*, 1055–1073. [CrossRef]
35. Geisler, J.H.; Colbert, M.W.; Carew, J.L. A new fossil species supports and early origin for toothed whale echolocation. *Nature* **2014**, *508*, 383–386. [CrossRef]
36. Mourlam, M.J.; Orliac, M.J. Infrasonic and ultrasonic hearing evolved after the emergence of modern whales. *Curr. Biol.* **2017**, *27*, 1776–1781. [CrossRef]
37. Pilleri, G. Adaptation to water and the evolution of echolocation in the Cetacea. *Ethol. Ecol. Evol.* **1990**, *2*, 135–163. [CrossRef]
38. Ketten, D.R. The marine mammal ear: Specializations for aquatic audition and echolocation. In *The Evolutionary Biology of Hearing*; Webster, D.B., Popper, A.N., Fay, R.R., Eds.; Springer: New York, NY, USA, 1992; pp. 717–750.
39. Werth, A.J. Adaptations of the cetacean hyolingual apparatus for aquatic feeding and thermoregulation. *Anat. Rec.* **2007**, *290*, 546–568. [CrossRef]
40. Fasick, J.I.; Robinson, P.R. Adaptations of cetacean retinal pigments to aquatic environments. *Front. Ecol. Evol.* **2016**, *4*, e70. [CrossRef]
41. Noren, S.R.; Williams, T.M. Body size and skeletal muscle myoglobin of cetaceans: Adaptations for maximizing dive duration. *Comp. Biochem. Physiol. A* **2000**, *126*, 181–191. [CrossRef]
42. McClellan, D.A.; Palfreyman, E.J.; Smith, M.J.; Moss, J.L.; Christensen, R.G.; Sailsbery, J.K. Physicochemical evolution and molecular adaptation of the cetacean and artiodactyl cytochrome b proteins. *Mol. Biol. Evol.* **2005**, *22*, 437–455. [CrossRef] [PubMed]
43. Shen, T.; Xu, S.; Wang, X.; Yu, W.; Zhou, K.; Yang, G. Adaptive evolution and functional constraint at TLR4 during the secondary aquatic adaptation and diversification of cetaceans. *BMC Evol. Biol.* **2012**, *12*, e39. [CrossRef]

44. Slater, G.J.; Price, S.A.; Santini, F.; Alfaro, M.E. Diversity versus disparity and the radiation of modern cetaceans. *Proc. R. Soc. B* **2010**, *277*, 3097–3104. [CrossRef]
45. Steeman, M.E.; Hebagaard, M.B.; Fordyce, R.E.; Ho, S.Y.W.; Rabosky, D.L.; Nielsen, R.; Rahbek, C.; Glenner, H.; Sorensen, M.V.; Willersley, E. Radiation of extant cetaceans driven by restructuring of the oceans. *Syst. Biol.* **2009**, *58*, 573–585. [CrossRef]
46. Marx, F.G.; Fordyce, R.E. Baleen boom and bust: A synthesis of mysticete phylogeny, diversity, and disparity. *R. Soc. Open Sci.* **2015**, *2*, e140434. [CrossRef]
47. Sacher, G.A. Constitutional basis of longevity in the Cetacea: Do the whales and the terrestrial mammals obey the same laws? *Rep. Int. Whal. Comm. Spec. Issue* **1980**, *3*, 209–213.
48. Seim, I.; Ma, S.; Zhou, X.; Gerashchenko, M.V.; Lee, S.G.; Suydam, R.; George, J.C.; Bickham, J.W.; Gladyshev, V.N. The transcriptome of the bowhead whale Balaena mysticetus reveals adaptations of the longest-lived mammal. *Aging* **2014**, *6*, 879–899. [CrossRef] [PubMed]
49. Fertl, D.; Rosen, P.E. Albinism. In *Encyclopedia of Marine Mammals*, 3rd ed.; Würsig, B., Thewissen, J.G.M., Kovacs, K.M., Eds.; Academic Press: San Diego, CA, USA, 2018; pp. 20–21.
50. Gero, S.; Englhaupt, D.; Rendell, L.; Whitehead, H. Who Cares? Between-group variation in alloparental caregiving in sperm whales. *Behav. Ecol.* **2009**, *20*, 838–843. [CrossRef]
51. Augusto, J.F.; Frasier, T.R.; Whitehead, H. Characterizing alloparental care in the pilot whale (*Globicephala melas*) population that summers off Cape Breton, Nova Scotia, Canada. *Mar. Mamm. Sci.* **2016**, *33*, 440–456. [CrossRef]
52. Weinpress, M.R.; Herzing, D. Maternal and Alloparental Discipline in Atlantic Spotted Dolphins (Stenella frontalis) in the Bahamas. *Anim. Behav. Cogn.* **2015**, *2*, 348–364. [CrossRef]
53. Hare, M.P.; Cipriano, F.; Palumbi, S.R. Genetic evidence on the demography of speciation in allopatric dolphin species. *Evolution* **2002**, *56*, 804–816. [CrossRef]
54. Connor, R.; Norris, K.S. Are Dolphins Reciprocal Altruists? *Am. Nat.* **1982**, *119*, 358–374. [CrossRef]
55. Pitman, R.L.; Deecke, V.B.; Gabriele, C.M.; Srinivasan, M.; Black, N.; Denkinger, J.; Durban, J.W.; Matthews, E.A.; Matkin, D.R.; Neilson, J.L.; et al. Humpback whales interfering when mammal-eating killer whales attack other species: Mobbing behavior and interspecific altruism? *Mar. Mamm. Sci.* **2017**, *33*, 7–58. [CrossRef]
56. Fordyce, R.E. Cetacean evolution. In *Encyclopedia of Marine Mammals*, 3rd ed.; Würsig, B., Thewissen, J.G.M., Kovacs, K.M., Eds.; Academic Press: San Diego, CA, USA, 2018; pp. 180–185.
57. Pyenson, N.D.; Sponberg, S.N. Reconstructing body size in extinct crown Cetacea (Neoceti) using allometry, phylogenetic methods and tests from the fossil record. *J. Mamm. Evol.* **2011**, *18*, e269. [CrossRef]
58. Didier, G. Time-dependent-asymmetric-linear-parsimonious ancestral state reconstruction. *Bull. Math. Biol.* **2017**, *79*, 2334–2355. [CrossRef]
59. Johnston, C.; Berta, A. Comparative anatomy and evolutionary history of suction feeding in cetaceans. *Mar. Mamm. Sci.* **2011**, *27*, 493–513. [CrossRef]
60. Geisler, J.H.; Luo, Z. The petrosal and inner ear of *Herpetocetus* sp. (Mammalia: Cetacea) and their implications for the phylogeny and hearing of archaic mysticetes. *J. Paleont.* **1996**, *70*, 1045–1066. [CrossRef]
61. Luo, Z. Homology and transformation of cetacean ectotympanic structures. In *The Emergence of Whales: Advances in Vertebrate Paleobiology*; Thewissen, J.G.M., Ed.; Springer: Boston, MA, USA, 1998; Volume 1, pp. 269–301.
62. O'Leary, M.A. Phylogenetic and morphometric reassessment of the dental evidence for a mesonychian and cetacean clade. In *The Emergence of Whales: Advances in Vertebrate Paleobiology*; Thewissen, J.G.M., Ed.; Springer: Boston, MA, USA, 1998; Volume 1, pp. 133–161.
63. Simões-Lopes, P.C.; Gutstein, C.S. Notes on the anatomy, positioning and homology of the pelvic bones in small cetaceans (Cetacea, Delphinidae, Pontoporiidae). *Lat. Am. J. Aq. Mamm.* **2004**, *3*, 157–162. [CrossRef]
64. Davies, J.L. The antitropical factor in cetacean speciation. *Evolution* **1963**, *17*, 107–116. [CrossRef]
65. Hamilton, H.; Caballero, S.; Collins, A.G.; Brownell, R.L. Evolution of river dolphins. *Proc. R. Soc. Lond. B* **2001**, *268*, 549–556. [CrossRef] [PubMed]
66. Loseto, L.L.; Stern, G.A.; Ferguson, S.H. Size and biomagnification: How habitat selection explains beluga mercury levels. *Environ. Sci. Technol.* **2008**, *42*, 3982–3988. [CrossRef] [PubMed]

67. Hoekstra, P.F.; O'Hara, T.M.; Pallant, S.J.; Solomon, K.R.; Muir, D.C. Bioaccumulation of Organochlorine Contaminants in Bowhead Whales (*Balaena mysticetus*) from Barrow, Alaska. *Arch. Environ. Contam. Toxicol.* **2002**, *42*, 497–507. [CrossRef]
68. Baron, E.; Gimenez, J.; Verborgh, P.; Gauffier, P.; DeStephanis, R.; Eljarrat, E.; Barcelo, D. Bioaccumulation and biomagnification of classical flame retardants, related halogenated natural compounds and alternative flame retardants in three delphinids from Southern European waters. *Environ. Pollut.* **2015**, *203*, 107–115. [CrossRef]
69. Alonso, M.B.; Azevedo, A.; Torres, J.P.; Dorneles, P.R.; Eljarrat, E.; Barcelo, D.; Lailson-Brito, J.; Malm, O. Anthropogenic (PBDE) and naturally-produced (MeO-PBDE) brominated compounds in cetaceans—A review. *Sci. Total Environ.* **2014**, *481*, 619–634. [CrossRef]
70. Bianucci, G.; Sart, G.; Catanzariti, R.; Santini, U. Middle Pliocene cetaceans from Monte Voltraio (Tuscany, Italy) biostratigraphical, paleoecological, and paleoclimatic observations. *Rev. Ital. Paleont. Strat.* **1998**, *104*, 1.
71. Pyenson, N.D.; Irmis, R.B.; Lipps, J.H.; Barnes, L.G.; Mitchell, E.D.; MacLeod, S.A. Origin of a widespread marine bonebed deposited during the middle Miocene Climatic Optimum. *Geology* **2009**, *37*, 519–522. [CrossRef]
72. Borrell, A.; Gomez-Campos, E.; Aguilar, A. Influence of reproduction on stable-isotope rations: Nitrogen and carbon isotope discrimination between mothers, fetuses, and milk in the fin whale, a capital breeder. *Physiol. Biochem. Zool.* **2016**, *89*, e684632. [CrossRef] [PubMed]
73. Christiansen, F.; Vikingsson, G.A.; Rasmussen, M.H.; Lusseau, D. Female body condition affects foetal growth in a capital breeding mysticete. *Funct. Ecol.* **2014**, *28*, 579–588. [CrossRef]
74. Roman, J.; McCarthy, J.J. The whale pump: Marine mammals enhance primary productivity in a coastal basin. *PLoS ONE* **2010**, *5*, e13255. [CrossRef] [PubMed]
75. Lavery, T.J.; Roudnew, B.; Gill, P.; Seymour, J.; Seuront, L.; Johnson, G.; Mitchell, J.G.; Smetacek, V. Iron defecation by sperm whales stimulates carbon export in the Southern Ocean. *Proc. R. Soc. Lond. B* **2010**, *277*, 3527–3531. [CrossRef]
76. Werth, A.J. Odontocete suction feeding: Experimental analysis of water flow and head shape. *J. Morph.* **2006**, *267*, 1415–1428. [CrossRef]
77. Werth, A.J. Mandibular and dental variation and the evolution of suction feeding in Odontoceti. *J. Mamm.* **2006**, *87*, 579–588. [CrossRef]
78. Bisconti, M. Taxonomy and evolution of the Italian Pliocene Mysticeti (Mammalia, Cetacea): A state of the art. *Bull. Soc. Paleont. Ital.* **2009**, *48*, 147–156.
79. Mead, J.G.; Mitchell, E.D. Atlantic gray whales. In *The Gray Whale: Eschrichtius robustus*; Jones, M.L., Swartz, S.L., Leatherwood, S., Eds.; Academic Press: San Diego, CA, USA, 1984; pp. 33–53.
80. Balbuena, J.A.; Raga, J.A. Ecology and host relationships of the whale-louse *Isocyamus delphini* (Amphipoda: Cyamidae) parasitizing long-finned pilot whales (*Globicephala melas*) off the Faroe Islands (Northeast Atlantic). *Can. J. Zool.* **1991**, *69*, 141–145. [CrossRef]
81. Mitchell, E. Pigmentation pattern evolution in delphinid cetaceans: An essay in adaptive coloration. *Can. J. Zool.* **1970**, *48*, 717–740. [CrossRef]
82. MacLeod, C.D. The relationship between body mass and relative investment in testes mass in cetaceans: Implications for inferring interspecific variations in the extent of sperm competition. *Mar. Mamm. Sci.* **2010**, *26*, 370–380. [CrossRef]
83. Ainley, D.G.; Ballard, G.; Dugger, K.M. Competition among penguins and cetaceans reveals trophic cascades in the Western Ross Sea, Antarctica. *Ecology* **2006**, *87*, 2080–2093. [CrossRef]
84. Collin, R.; Janis, C.M. Morphological constraints on tetrapod feeding mechanisms: Why were there no suspension-feeding marine reptiles? In *Ancient Marine Reptiles*; Callaway, J.M., Nicholls, E.L., Eds.; Academic Press: San Diego, CA, USA, 1997; pp. 451–466.
85. O'Keefe, F.R.; Otero, R.A.; Soto-Acuna, S.; O'Gorman, J.P.; Godfrey, S.J.; Chatterjee, S. Cranial anatomy of *Morturnia seymourensis* from Antarctica, and the evolution of filter feeding in plesiosaurts of the Austral Late Cretaceous. *J. Vert. Paleontol.* **2017**, *37*, e1347570. [CrossRef]
86. Muizon, C. Walrus-like feeding adaptation in a new cetacean from the Pliocene of Peru. *Nature* **1993**, *365*, 745–748. [CrossRef]
87. Muizon, C.; Domning, D.P.; Ketten, D.R. *Odobenocetops peruvianus*, the walrus-convergent delphinoid (Mammalia: Cetacea) from the early Pliocene of Peru. *Smithson. Contrib. Paleobiol.* **2002**, *93*, 223–261.

88. Fordyce, R.E.; Quilty, P.G.; Daniels, J. *Australodelphis mirus*, a bizarre new toothless ziphiid-like fossil dolphin (Cetacea: Delphinidae) from the Pliocene of Vestfold Hills, East Antarctica. *Antarct. Sci.* **2002**, *14*, 37–54. [CrossRef]
89. Ballance, L.T. Cetacean ecology. In *Encyclopedia of Marine Mammals*, 3rd ed.; Würsig, B., Thewissen, J.G.M., Kovacs, K.M., Eds.; Academic Press: San Diego, CA, USA, 2018; pp. 172–180.
90. Caro, T.; Beeman, K.; Stankowich, T.; Whitehead, H. The functional significance of colouration in cetaceans. *Evol. Ecol.* **2011**, *25*, e1231. [CrossRef]
91. Whitehead, H. Gene–culture coevolution in whales and dolphins. *Proc. Natl. Acad. Sci. USA* **2017**, *114*, 7814–7821. [CrossRef]
92. Rendell, L.; Whitehead, H. Culture in whales and dolphins. *Behav. Brain Sci.* **2001**, *24*, 309–382. [CrossRef]
93. Bearzi, G.; Eddy, L.; Piwetz, S.; Reggente, M.A.L.; Cozzi, B. Cetacean behavior toward the dead and dying. In *Encyclopedia of Animal Cognition and Behavior*; Vonk, J., Shackelford, T., Eds.; Springer Nature: Basel, Switzerland, 2017. [CrossRef]
94. Ishiyama, M. Enamel structure in odontocete whales. *Scanning Microsc.* **1987**, *1*, 1071–1079.
95. Werth, A.J.; Loch, C.; Fordyce, R.E. Enamel microstructure in Cetacea: A case study in evolutionary loss of complexity. *J. Mammal Evol.* **2019**. [CrossRef]
96. Johnstone, R.A.; Cant, M.A. The evolution of menopause in cetaceans and humans: The role of demography. *Proc. R. Soc. Lond. B* **2010**, *277*, 3765–3771. [CrossRef] [PubMed]
97. Gladden, J.G.B.; Ferguson, M.M.; Clayton, J.W. Matriarchal genetic population structure of North American beluga whales *Delphinapterus leucas* (Cetacea: Monodontidae). *Mol. Ecol.* **1997**, *6*, 1033–1046. [CrossRef] [PubMed]
98. Fontaine, M.C.; Tolley, K.A.; Michaux, J.R.; Birkun, A.; Ferreira, M.; Jauniaux, T.; Llavona, A.; Ozturk, B.; Ozturk, A.A.; Ridoux, V.; et al. Genetic and historic evidence for climate-driven population fragmentation in a top cetacean predator: The harbour porpoises in European water. *Proc. R. Soc. B* **2010**, *277*, 2829–2837. [CrossRef] [PubMed]
99. Lopes-Marques, M.; Machado, A.M.; Barbosa, S.; Fonseca, M.M.; Ruiyo, R.; Castro, L.F.C. Cetacea are natural knockouts for IL20. *Immunogenetics* **2018**, *70*, 681–687. [CrossRef] [PubMed]
100. Foote, A.D.; Vijay, N.; Avila-Arcos, M.C.; Baird, R.; Durban, J.W.; Fumagalli, M.; Gibbs, R.A.; Hanson, M.B.; Korneliussen, T.S.; Martin, M.D.; et al. Genome-culture coevolution promotes rapid divergence of killer whale ecotypes. *Nat. Commun.* **2016**, *7*, e11693. [CrossRef] [PubMed]
101. DeBruyn, P.J.; Tosh, C.A.; Terauds, A. Killer whale ecotypes: Is there a global model. *Biol. Rev. Camb. Philos. Soc.* **2013**, *88*, 62–80. [CrossRef]
102. Thewissen, J.G.M.; Heyning, J. Embryogenesis and development in *Stenella attenuata* and other cetaceans. In *Reproductive Biology and Phylogeny of Cetacea*; Miller, D.L., Ed.; CRC Press: Boca Raton, FL, USA, 2007; pp. 307–329.
103. Thewissen, J.G.M.; Hieronymus, T.L.; George, J.C.; Suydam, R.; Stimmelmayr, R.; McBurney, D. Evolutionary aspects of the development of teeth and baleen in the bowhead whale. *J. Anat.* **2017**, *230*, 549–566. [CrossRef]
104. Marino, L. A comparison of encephalization between odontocete cetaceans and anthropoid primates. *Brain. Behav. Evol.* **1998**, *51*, 230–238. [CrossRef]
105. Perez-Alvarez, M.J.; Olavarria, C.; Moraga, R.; Baker, C.S.; Hamner, R.M.; Poulin, E. Microsatellite markers reveal strong genetic structure in the endemic Chilean dolphin. *PLoS ONE* **2015**, *10*, e0123956. [CrossRef]
106. Beal, A.P.; Kiszka, J.J.; Wells, R.S.; Eirin-Lopez, J.M. The bottlenose dolphin epigenetic aging tool (BEAT): A molecular age estimation tool for small cetaceans. *Front. Mar. Sci.* **2019**, *6*, e561. [CrossRef]
107. Tejada-Martinez, D.; Magalhaes, J.P.; Opazo, J.C. Positive selection and fast turnover rate in tumor suppressor genes reveal how cetaceans resist cancer. *bioRxiv* **2020**. [CrossRef]
108. Bearzi, G.; Reggente, M.A.L. Epimeletic behavior. In *Encyclopedia of Marine Mammals*, 3rd ed.; Würsig, B., Thewissen, J.G.M., Kovacs, K.M., Eds.; Academic Press: San Diego, CA, USA, 2018; pp. 337–338.
109. Caldwell, M.C.; Caldwell, D.K. Epimeletic (care-giving) behavior in Cetacea. In *Whales, Dolphins, and Porpoises*; Norris, K.N., Ed.; University of California Press: Berkeley, CA, USA, 1966; pp. 755–789.
110. Berta, A.; Ekdale, E.G.; Cranford, T.W. Review of the cetacean nose: Form, function, and evolution. *Anat. Rec.* **2014**, *297*, 2205–2215. [CrossRef] [PubMed]

111. Cranford, T.W. The sperm whale's nose: Sexual selection on a grand scale? *Mar. Mamm. Sci.* **1999**, *15*, 1133–1157. [CrossRef]
112. Marx, F.G.; Fitzgerald, E.M.G.; Fordyce, R.E. Like phoenix from the ashes: How modern baleen whales arose from a fossil "dark age.". *Acta Palaeontol. Pol.* **2019**, *64*, 231–238. [CrossRef]
113. Armfield, B.A.; Zheng, Z.; Bajpai, S.; Vinyard, C.J.; Thewissen, J.G.M. Development and evolution of the unique cetacean dentition. *PeerJ* **2013**, *1*, e24. [CrossRef]
114. Bejder, L.; Hall, B.K. Limbs in whales and limblessness in other vertebrates: Mechanisms of evolutionary and developmental transformation and loss. *Evol. Dev.* **2002**, *4*, 445–458. [CrossRef]
115. Fahlke, J.M.; Hampe, O. Cranial symmetry in baleen whales (Cetacea, Mysticeti) and the occurrence of cranial asymmetry throughout cetacean evolution. *Sci. Nat.* **2015**, *102*, e58. [CrossRef]
116. Lindberg, D.R.; Pyenson, N.D. Things that go bump in the night: Evolutionary interactions between cephalopods and cetaceans in the tertiary. *Lethaia* **2007**, *40*, 335–343. [CrossRef]
117. Geisler, J.H.; McGowen, M.R.; Yang, G.; Gatesy, J. A supermatrix analysis of genomic, morphological, and paleontological data from crown Cetacea. *BMC Evol. Biol.* **2011**, *11*, e112. [CrossRef]
118. Turvey, S.T.; Pitman, R.L.; Taylor, B.L.; Barlow, J.; Akamatsu, T.; Barrett, L.A.; Zhao, X.; Reeves, R.R.; Stewart, B.S.; Wang, K.; et al. First human-caused extinction of a cetacean species? *Biol. Lett.* **2007**, *3*, 537–540. [CrossRef]
119. Snively, E.; Fahlke, J.M.; Welsh, R.C. Bone-breaking bite force of *Basilosaurus isis* (Mammalia, Cetacea) from the Late Eocene of Egypt estimated by finite element analysis. *PLoS ONE* **2015**, *10*, e0118380. [CrossRef] [PubMed]
120. Parra, G.J.; Corkeron, P.J.; Arnold, P. Grouping and fission–fusion dynamics in Australian snubfin and Indo-Pacific humpback dolphins. *Anim. Behav.* **2011**, *82*, 1423–1433. [CrossRef]
121. Best, R.C. The tusk of the narwhal (*Monodon monoceros* L.): Interpretation of its function (Mammalia: Cetacea). *Can. J. Zool.* **1981**, *59*, 2386–2393. [CrossRef]
122. Nweeia, M.T.; Eichmiller, F.C.; Nutarak, C.; Eidelman, N.; Giuseppetti, A.A.; Quinn, J.; Mead, J.G.; K'issuk, K.; Hauschka, P.V.; Tyler, E.M. Considerations of anatomy, morphology, evolution, and function for narwhal dentition. In *Smithsonian at the Poles: Contributions to International Polar Year Science*; Lang, M.A., Miller, S.E., Eds.; Smithsonian Press: Washington, DC, USA, 2009; pp. 223–240.
123. Kiel, S.; Goedert, J.L. Deep-sea food bonanzas: Early Cenozoic whale-fall communities resemble wood-fall rather than seep communities. *Proc. R. Soc. Lond. B* **2006**, *273*, 2625–2632. [CrossRef] [PubMed]
124. Fujiwara, Y.; Kawato, M.; Yamanaka, T.; Sato-Okoshi, W.; Noda, C.; Tsuchida, S.; Komai, T.; Cubelio, S.S. Three-year investigations into sperm whale-fall ecosystems in Japan. *Mar. Ecol.* **2007**, *28*, 219–232. [CrossRef]
125. McGowen, M.R.; Spaulding, M.; Gatesy, J. Divergence date estimation and a comprehensive molecular tree of extant cetaceans. *Mol. Phylogen. Evol.* **2009**, *53*, 891–906. [CrossRef]
126. Bosio, G.; Malinverno, E.; Villa, I.M.; DiCelma, C.; Gariboldi, K.; Gioncada, A.; Barberini, V.; Urbina, M.; Bianucci, G. Tephrochronology and chronostratigraphy of the Miocene Chilcatay and Pisco formations (East Pisco Basin, Peru). *Newslett. Stratigr.* **2020**, *53*, 213–247. [CrossRef]
127. Gariboldi, K.; Bosio, G.; Malinverno, E.; Gioncada, A.; DiCelma, C.; Villa, I.M.; Urbina, M.; Bianucci, G. Biostratigraphy, geochronology, and sedimentation rates of the upper Miocene Pisco Formation at two important marine vertebrate fossil-bearing sites of southern Peru. *Newslett. Stratigr.* **2017**, *50*, 417–444. [CrossRef]
128. Gioncada, A.; Collareta, A.; Gariboldi, K.; Lambert, O.; DiCelma, C.; Bonaccorsi, E.; Urbina, M.; Bianucci, G. Inside baleen: Exceptional microstructure preservation in a late Miocene whale skeleton from Peru. *Geology* **2016**, *44*, 839–842. [CrossRef]
129. Brand, L.; Urbina, M.; Chadwick, A.; DeVries, T.J.; Esperante, R. A high resolution stratigraphic framework for the remarkable fossil cetacean assemblage of the Miocene/Pliocene Pisco Formation, Peru. *J. S. Am. Earth Sci.* **2011**, *31*, 414–425. [CrossRef]
130. Marino, L.; Uhen, M.D.; Pyenson, N.D.; Frohlich, B. Reconstructing cetacean brain evolution using computed tomography. *Anat. Rec.* **2003**, *272B*, 107–117. [CrossRef] [PubMed]
131. Buchholtz, E.A. Vertebral osteology and swimming style in living and fossil whales (Order: Cetacea). *J. Zool.* **2001**, *253*, 175–190. [CrossRef]

132. Lee, K.; Lee, J.M.; Sohn, S.; Cho, Y.; Choi, Y.M.; Kim, H.K.; Kim, J.H.; Jeong, D.G. Complete mitochondrial genome of the Pacific white-sided dolphin *Lagenorhynchus obliquidens* (Cetacea: Delphinidae). *Conserv. Gen. Res.* **2018**, *10*, 201–204. [CrossRef]
133. Rooney, A.; Honeycutt, R.; Davis, S.; Derr, J.N. Evaluating a putative bottleneck in a population of bowhead whales from patterns of microsatellite diversity and genetic disequilibria. *J. Mol. Evol.* **1999**, *49*, 682–690. [CrossRef]
134. Amendola-Pimenta, M.; Camelo-Marrufo, M.; Zamora-Briseno, J.A.; Hernandez-Velazques, I.M.; Zamora-Bustillos, R.; Rodrigues-Canul, R. Genetic bottleneck and founder effect signatures in a captive population of common bottlenose dolphins *Tursiops truncatus* (Montagu 1821) in Mexico. *PeerJ* **2018**, *6*, e26891v1.
135. Gaspari, S.; Airoldi, S.; Hoelzel, A.R. Risso's dolphins (*Grampus griseus*) in UK waters are differentiated from a population in the Mediterranean Sea and genetically less diverse. *Conserv. Gen.* **2007**, *8*, 727–732. [CrossRef]
136. McGowan, M.R.; Tsagkogeorga, G.; Alvarez-Carretero, S.; dos Reis, M.; Struebig, M.; Deaville, R.; Jepson, P.D.; Jarman, S.; Polanowski, A.; Morin, P.A. Phylogenomic resolution of the cetacean tree of life using target sequence capture. *Syst. Biol.* **2020**, *69*, 479–501. [CrossRef]
137. Nery, M.F.; Gonzalez, D.J.; Opazo, J.C. How to Make a Dolphin: Molecular Signature of Positive Selection in Cetacean Genome. *PLoS ONE* **2013**, *8*, e65491. [CrossRef]
138. Yim, H.S.; Cho, Y.S.; Guang, X.; Kang, S.G.; Jeong, J.Y.; Cha, S.S.; Oh, H.M.; Lee, J.H.; Yang, E.C.; Kwon, K.K.; et al. Minke whale genome and aquatic adaptation in cetaceans. *Nat. Gen.* **2014**, *46*, 88–92. [CrossRef]
139. Goldbogen, J.A.; Madsen, P.T. The evolution of foraging capacity and gigantism in cetaceans. *J. Exp. Biol.* **2018**, *221*, e166033. [CrossRef]
140. Fordyce, R.E.; Marx, F.G. Gigantism precedes filter feeding in baleen whale evolution. *Curr. Biol.* **2018**, *28*, 1670–1676. [CrossRef] [PubMed]
141. Bianucci, G.; Marx, F.G.; Collareta, A.; DiStefano, A.; Landini, W.; Morigi, C.; Varola, A. Rise of the titans: Baleen whales became giants earlier than thought. *Biol. Lett.* **2019**, *15*, e0175. [CrossRef] [PubMed]
142. Lambert, O.; Bianucci, G.; Post, K.; de Muizon, C.; Salas-Gismondi, R.; Urbina, M.; Reuner, J. The giant bite of a new raptorial sperm whale from the Miocene epoch of Peru. *Nature* **2010**, *466*, 105–108. [CrossRef] [PubMed]
143. Whitehead, H. Cultural selection and genetic diversity in matrilineal whales. *Science* **1998**, *28*, 1708–1711. [CrossRef] [PubMed]
144. Huang, S.L.; Hao, Y.; Mei, Z.; Turvey, S.T.; Wang, D. Common pattern of population decline for freshwater cetacean species in deteriorating habitats. *Freshw. Biol.* **2012**, *57*, 1266–1276. [CrossRef]
145. Gol'din, P. Naming an innominate: Pelvis and hindlimbs of Miocene whales give and insight into evolution and homology of cetacean pelvic girdle. *Evol. Biol.* **2017**, *41*, 473–479. [CrossRef]
146. Peredo, C.M.; Pyenson, N.D.; Uhen, M.D.; Marshall, C.D. Alveoli, teeth, and tooth loss: Understanding the homology of internal mandibular structures in mysticete cetaceans. *PLoS ONE* **2017**, *12*, e0178243. [CrossRef]
147. Fraija-Fernandez, N.; Olson, P.D.; Crespo, E.A.; Raga, J.A.; Aznar, F.J.; Fernandez, M. Independent host switching events by digenean parasites of cetaceans inferred from ribosomal DNA. *Int. J. Parasitol.* **2015**, *45*, 167–173. [CrossRef]
148. Marx, F.G.; Uhen, M.D. Climate, critters, and cetaceans: Cenozoic drivers of the evolution of modern whales. *Science* **2010**, *327*, 993–996. [CrossRef]
149. Simões-Lopes, P.C.; Daura-Jorge, F.G.; Cantor, M. Clues of cultural transmission in cooperative foraging between artisanal fishermen and bottlenose dolphins, *Tursiops truncatus* (Cetacea: Delphinidae). *Zool. Curitiba* **2016**, *33*. [CrossRef]
150. Peterson, D.; Hanazaki, N.; Simões-Lopes, P.C. Natural resource appropriation in cooperative artisanal fishing between fishermen and dolphins (*Tursiops truncatus*) in Laguna, Brazil. *Ocean Coast. Manag.* **2008**, *51*, 469–475. [CrossRef]
151. Crossman, C.A.; Taylor, E.B.; Barrett-Lennard, L.G. Hybridization in the Cetacea: Widespread occurrence and associated morphological, behavioral, and ecological factors. *Ecol. Evol.* **2016**, *6*, 1293–1303. [CrossRef] [PubMed]

152. Van Bressem, M.F.; Raga, J.A.; DiGuardo, G.; Jepson, P.D.; Duignan, P.J.; Siebert, U.; Barrett, T.; Santos, M.C.O.; Moreno, I.B.; Siciliano, S.; et al. Emerging infectious diseases in cetaceans worldwide and the possible role of environmental stressors. *Dis. Aquat. Org.* **2009**, *86*, 143–157. [CrossRef] [PubMed]
153. Darwin, C.R. *On the Origin of Species*; John Murray: London, UK, 1859.
154. Gillet, A.; Frederich, B.; Parmentier, E. Divergent evolutionary morphology of the axial skeleton as a potential key innovation in modern cetaceans. *Proc. R. Soc. B* **2019**, *286*, e20191771. [CrossRef] [PubMed]
155. Estes, J.A.; Tinker, M.T.; Williams, T.M.; Doak, D.F. Killer whale predation on sea otters linking oceanic and nearshore ecosystems. *Science* **1998**, *285*, 473–476. [CrossRef]
156. Parsons, K.M.; Durban, J.W.; Claridge, D.E.; Balcomb, K.C.; Noble, L.R.; Thompson, P.M. Kinship as a basis for alliance formation between male bottlenose dolphins, Tursiops truncatus, in the Bahamas. *Anim. Behav.* **2003**, *66*, 185–194. [CrossRef]
157. Huang, S.L.; Ni, I.H.; Chou, L.S. Correlations in cetacean life history traits. *Raffles Bull. Zool.* **2008**, *19*, 285–292.
158. Marx, F.G.; Fordyce, R.E. A link no longer missing: New evidence for the cetotheriid affinity for *Caperea*. *PLoS ONE* **2016**, *11*, e0164059. [CrossRef]
159. Boessenecker, R.W. Pleistocene survival of an archaic dwarf baleen whale (Mysticeti: Cetotheriidae). *Naturwissenschaften* **2013**, *100*, 365–371. [CrossRef]
160. Perrin, W.M.; Mesnick, S.L. Sexual ecology of the spinner dolphin, *Stenella longirostris*: Geographic variation in mating system. *Mar. Mamm. Sci.* **2003**, *19*, 462–483. [CrossRef]
161. Schaeff, C.M. Courtship and mating behavior. In *Reproductive Biology and Phylogeny of Cetacea*; Miller, D.L., Ed.; CRC Press: Boca Raton, FL, USA, 2007; pp. 349–370.
162. Hershkovitz, P. *Catalogue of Living Whales*; Smithsonian: Washington, DC, USA, 1966.
163. Corkeron, P.J.; Connor, R.C. Why do baleen whales migrate? *Mar. Mamm. Sci.* **2006**, *15*, 1228–1245. [CrossRef]
164. Churchill, M.; Miguel, J.; Beatty, B.L.; Gowami, A.; Geisler, J.H. Asymmetry drives modularity of the skull in the common dolphin (Delphinus delphis). *Biol. J. Linn. Soc.* **2018**, *126*, 225–239. [CrossRef]
165. Jackson, J.A.; Baker, C.S.; Vant, M.; Steel, D.J.; Medrano-Gonzalez, L.; Palumbi, S.R. Big and slow: Phylogenetic estimates of molecular evolution in baleen whales (Suborder Mysticeti). *Mol. Biol. Evol.* **2009**, *26*, 2427–2440. [CrossRef] [PubMed]
166. Milinkovitch, M.C.; Mayer, A.; Powell, J.R. Phylogeny of all major groups of cetaceans based on DNA sequences from three mitochondrial genes. *Mol. Biol. Evol.* **1994**, *11*, 939–948. [PubMed]
167. Theodor, J.M. Molecular clock divergence estimates and the fossil record of Cetartiodactyla. *J. Paleont* **2004**, *78*, 39–44. [CrossRef]
168. Naylor, G.J.P.; Adams, D.C. Are the fossil data really at odds with the molecular data? Morphological evidence for Cetartiodactyl phylogeny reexamined. *Syst. Biol.* **2001**, *50*, 444–453.
169. Messenger, S.L.; McGuire, J.A. Morphology, molecules, and the phylogenetics of cetaceans. *Syst. Biol.* **1998**, *47*, 90–124. [CrossRef]
170. Amaral, A.R.; Coelho, M.M.; Marugan-Lobon, J.; Rohlf, F.J. Cranial shape differentiation in three closely related delphinid cetacean species: Insights into evolutionary history. *Zoology* **2009**, *112*, 38–47. [CrossRef]
171. Falk, D.; Dudek, B. Mosaic evolution of the neocortex. *Behav. Brain Sci.* **1993**, *16*, 701–702. [CrossRef]
172. Zhu, K.; Zhou, X.; Xu, S.; Sun, D.; Ren, W.; Zhou, K.; Yang, G. The loss of taste genes in cetaceans. *BMC Evol. Biol.* **2014**, *14*, e218. [CrossRef]
173. Huelsmann, M.; Hecker, N.; Springer, M.S.; Gatesy, J.; Sharma, V.; Hiller, M. Genes lost during the transition from land to water in cetaceans highlight genomic changes associated with aquatic adaptations. *Sci. Adv.* **2019**, *5*, e6671. [CrossRef] [PubMed]
174. Chen, Z.; Wang, Z.; Xu, S.; Zhou, K.; Yang, G. Characterization of hairless (*Hr*) and *FGF5* genes provides insights into the molecular basis of hair loss in cetaceans. *BMC Evol. Biol.* **2013**, *13*, e34. [CrossRef] [PubMed]
175. Katona, S.; Whitehead, H. Are Cetacea ecologically important? *Oceanogr. Mar. Biol. Ann. Rev.* **1988**, *26*, 553–568.
176. Trull, P. Symbiotic Relationship between Humpback Whales and Marine Birds. Orleans Conservation Trust. Available online: https://orleansconservationtrust.org/symbiotic-relationship-between-humpback-whales-and-marine-birds-presentation-recap/ (accessed on 1 April 2020).
177. Tsai, C.H.; Fordyce, R.E. Juvenile morphology in baleen whale phylogeny. *Naturwissenschaften* **2014**, *101*, 765–769. [CrossRef]

178. Galatius, A. Paedomorphosis in two small species of toothed whales (Odontoceti): How and why? *Biol. J. Linn. Soc.* **2010**, *99*, 278–295. [CrossRef]
179. Waples, R.S. Genetic methods for estimating the effective size of cetacean populations. *Rep. Int. Whal. Comm.* **1991**, *13*, 279–300.
180. Friedlaender, A.; Lawon, G.L.; Halpin, P.N. Evidence of resource partitioning between humpback and minke whales around the western Antarctic Peninsula. *Mar. Mamm. Sci.* **2009**, *25*, 402–415. [CrossRef]
181. Erwin, D.H. Macroevolution: Dynamics of diversity. *Curr. Biol.* **2011**, *21*, R1000–R1001. [CrossRef]
182. Milinkovitch, M.C.; Berube, M.; Palsboll, P.J. Cetaceans are highly derived artiodactyls. In *The Emergence of Whales*; Thewissen, J.G.M., Ed.; Plenum: New York, NY, USA, 1998; pp. 113–131.
183. Santos, M.; Rosso, S. Ecological aspects of marine tucuxi dolphins (*Sotalia guianensis*) based on group size and composition in the Cananéia Estuary, southeastern Brazil. *Lat. Am. J. Aquat. Mamm.* **2007**, *6*, 71–82. [CrossRef]
184. Collareta, A.; Lambert, O.; Landini, W.; DiCelma, C.; Malinverno, E.; Varas-Malca, R.; Urbina, M.; Bianucci, G. Did the giant extinct shark *Carcharocles megalodon* target small prey? Bite marks on marine mammal remains from the late Miocene of Peru. *Paleogeog. Paleoclimatol. Paleoecol.* **2017**, *469*, 84–91. [CrossRef]
185. Yablokov, A.V. Convergence or parallelism in the evolution of cetaceans. *Int. Geol. Rev.* **1965**, *7*, 1461–1468. [CrossRef]
186. Perrin, W.F.; Rosel, P.E.; Cipriano, F. How to contend with paraphyly in the taxonomy of delphinine cetaceans? *Mar. Mamm. Sci.* **2013**, *29*, 567–588. [CrossRef]
187. Arvy, L. Phoresies and parasitism in cetaceans: A review. *Invest Cetacea* **1982**, *14*, 233–335.
188. Paulos, R.D.; Trone, M.; Kuczaj, S.A. Play in wild and captive cetaceans. *Int. J. Comp. Psychol.* **2010**, *23*, 701–722.
189. Wang, J.Y.; Liao, W.B. Ontogenesis and evolutionary allometry shape divergent evolution of genitalia in female cetaceans. *Evolution* **2018**, *72*, 404–405. [CrossRef] [PubMed]
190. Loch, C.; Duncan, W.; Simões-Lopes, P.C.; Kieser, J.A.; Fordyce, R.E. Ultrastructure of enamel and dentine in extant dolphins (Cetacea: Delphinoidea and Inioidea). *Zoomorphology* **2013**, *132*, 215–225. [CrossRef]
191. Cooper, L.N.; Berta, A.; Dawson, S.D.; Reidenberg, J.S. Evolution of hyperphalangy and digit reduction in the cetacean manus. *Anat. Rec.* **2007**, *290*, 654–672. [CrossRef]
192. Cercio, S.; Jacobsen, J.K.; Cholewiak, D.M.; Falcone, E.A.; Merriwether, D.A. Paternity in humpback whales, Megaptera novaeangliae: Assessing polygyny and skew in male reproductive success. *Anim. Behav.* **2005**, *70*, 267–277. [CrossRef]
193. Vachon, F.; Whitehead, H.; Frasier, T.R. What factors shape genetic diversity in cetaceans? *Ecol. Evol.* **2018**, *8*, 1554–1572. [CrossRef]
194. Hoelzel, A.R.; Hancock, J.M.; Dover, G.A. Evolution of the cetacean mitochondrial D-loop region. *Mol. Biol. Evol.* **1991**, *8*, 475–493.
195. Pagan, H.J.T.; Ferrer, T.; O'Corry-Crowe, G. Positive selection in coding regions and motif duplication in regulatory regions of bottlenose dolphin MHC class II genes. *PLoS ONE* **2018**, *13*, e0203450. [CrossRef] [PubMed]
196. Kerem, D.; Goffman, O.; Elasar, M.; Hadar, N.; Scheinin, A.; Lewis, T. The rough-toothed dolphin, *Steno bredanensis*, in the Eastern Mediterranean Sea: A relict population? *Adv. Mar. Biol.* **2016**, *75*, 233–258. [PubMed]
197. Riesch, R.; Barrett-Lennard, L.G.; Ellis, G.M.; Ford, J.K.B.; Deecke, V.B. Cultural traditions and the evolution of reproductive isolation: Ecological speciation in killer whales? *Biol. J. Linn. Soc.* **2012**, *106*, 1–17. [CrossRef]
198. Marsh, H.; Kasuya, T. 1986 Evidence for reproductive senescence in female cetaceans. *Rep. Int. Whal. Comm.* **1986**, *8*, 57–73.
199. Ellis, S.; Franks, D.W.; Nattrass, S.; Currie, T.E.; Cant, M.A.; Giles, D.; Balcomb, K.C.; Croft, D.P. Analyses of ovarian activity reveal repeated evolution of post-reproductive lifespans in toothed whales. *Sci. Rep.* **2018**, *8*, e12833. [CrossRef] [PubMed]
200. Ansmann, I.C.; Lanyon, J.M.; Seddon, J.M.; Parra, G.J. Habitat and resource partitioning among Indo-Pacific bottlenose dolphins in Moreton Bay, Australia. *Mar. Mamm. Sci.* **2015**, *31*, 211–230. [CrossRef]
201. Pyenson, N.D.; Goldbogen, J.A.; Shadwick, R.E. Mandible allometry in extant and fossil Balaenopteridae (Cetacea: Mammalia): The largest vertebrate skeletal element and its role in rorqual lunge feeding. *Biol. J. Linn. Soc.* **2013**, *108*, 586–599. [CrossRef]

202. Alexander, A.; Steel, D.; Slikas, B.; Hoekzema, K.; Carraher, C.; Parks, M.; Cronn, R.; Baker, C.S. Low diversity in the mitogenome of sperm whales revealed by next generation sequencing. *Genome Biol. Evol.* **2013**, *5*, 113–129. [CrossRef]
203. Zheng, R.; Karczmarki, L.; Lin, W.; Chan, S.C.Y.; Chang, W.L.; Wu, Y. Infanticide in the Indo-Pacific humpback dolphin (*Sousa chinensis*). *J. Ethol.* **2016**, *34*, 299–307. [CrossRef]
204. Dalebout, M.L.; Steel, D.; Baker, C.S. Phylogeny of the beaked whale genus Mesoplodon (Ziphiidae: Cetacea) revealed by nuclear introns: Implications for the evolution of male tusks. *Syst. Biol.* **2008**, *57*, 857–875. [CrossRef]
205. Borrell, A.; Velásquez Vacca, A.; Pinela, A.M.; Kinze, C.; Lockyer, C.H.; Vighi, M.; Aguilar, A. Stable isotopes provide insight into population structure and segregation in Eastern North Atlantic sperm whales. *PLoS ONE* **2013**, *8*, e82398. [CrossRef] [PubMed]
206. Clementz, M.T.; Fordyce, R.E.; Peek, S.L.; Fox, D.L. Ancient marine isoscapes and isotopic evidence of bulk-feeding by Oligocene cetaceans. *Paleogeogr. Paleoclimatol. Paleoecol.* **2014**, *400*, 28–40. [CrossRef]
207. Perrin, W.F.; Mead, J.G.; Brownell, R.L. Review of the Evidence Used in the Description of Currently Recognized Cetacean Subspecies. 2009. Available online: https://digitalcommons.unl.edu/cgi/viewcontent.cgi?article=1115&context=usdeptcommercepub (accessed on 21 May 2020).
208. Yang, G.; Liu, S.; Ren, W.; Zhou, K.; Wei, F. Mitochondrial control region variability of baiji and the Yangtze finless porpoises, two sympatric small cetaceans in the Yangtze river. *Acta Theriol.* **2003**, *48*, 469–483. [CrossRef]
209. Hoelzel, A.R. Genetic structure of cetacean populations in sympatry, parapatry, and mixed assemblages: Implications for conservation policy. *J. Hered.* **1998**, *89*, 451–458. [CrossRef]
210. Boessenecker, R.W.; Perry, F.A.; Schmitt, J.G. Comparative taphonomy, taphofacies, and bonebeds of the Mio-Pliocene Purisima Formation, Central California: Strong physical control on marine vertebrate preservation in shallow marine settings. *PLoS ONE* **2014**, *9*, e91419. [CrossRef] [PubMed]
211. Bianucci, G.; Collareta, A.; Bosio, G.; Landini, W.; Gariboldi, K.; Gioncada, A.; Lambert, O.; Malinverno, E.; de Muizon, C.; Varas-Malca, R.; et al. Taphonomy and palaeoecology of the lower Miocene marine vertebrate assemblage of Ullujaya (Chilcatay Formation, East Pisco Basin, southern Peru). *Paleogeogr. Paleoclimatol. Paleoecol.* **2018**, *511*, 256–279. [CrossRef]
212. Esperante, R.; Brand, L.; Chadwick, A.; Poma, O. Taphonomy of fossil whales in the diatomaceous sediments of the Miocene/Pliocene Pisco Formation, Peru. *Curr. Top. Taph. Fossil.* **2002**, 337–344.
213. Krutzen, M.; Mann, J.; Heithaus, M.R.; Connor, R.C.; Bajder, L.; Sherwin, W.B. Cultural transmission of tool use in bottlenose dolphins. *Proc. Natl. Acad. Sci. USA* **2005**, *102*, 8939–8943. [CrossRef]
214. Baum, J.K.; Worm, B. Cascading top-down effects of changing ocean predator abundances. *J. Anim. Ecol.* **2009**, *78*, 699–714. [CrossRef]
215. Lynam, C.P.; Llope, M.; Mollman, C.; Helaouet, P.; Bayliss-Brown, G.A.; Stenseth, N.C. Interaction between top-down and bottom-up control in marine food webs. *Proc. Natl. Acad. Sci. USA* **2017**, *114*, 1952–1957. [CrossRef]
216. Ainley, D.; Ballard, G.; Blight, L.K.; Ackley, S.; Emslie, S.D.; Lescroel, A.; Olmastroni, S.; Townsend, S.E.; Tynan, C.T.; Wilson, P.; et al. Impacts of cetaceans on the structure of Southern Ocean food webs. *Mar. Mamm. Sci.* **2010**, *26*, 482–498. [CrossRef]
217. Braulik, G.; Barnett, R.; Odon, V.; Islas-Villaneuva, V.; Hoelzel, R.; Graves, J.A. One species or two? Vicariance, lineage divergence, and low mtDNA diversity in geographically isolated populations of South Asian river dolphin. *J. Mamm. Evol.* **2014**, *22*, 111–120. [CrossRef]
218. Bisconti, M. Anatomy of a new cetotheriid genus and species from the Miocene of Herentals, Belgium, and the phylogenetic and palaeobiogeographical relationships of Cetotheriidae s.s. (Mammalia, Cetacea, Mysticeti). *J. Syst. Palaeont.* **2015**, *13*, 377–395. [CrossRef]

oceans MDPI

Article

Prey-Related Asphyxiation in Harbor Porpoises (*Phocoena phocoena*) along the U.S. West Coast: Importance of American Shad (*Alosa sapidissima*) on Adult Female Harbor Porpoise Mortality

Cindy R. Elliser [1,*], John Calambokidis [2], Dalin N. D'Alessandro [3], Deborah A. Duffield [3], Jessica L. Huggins [2], James Rice [4], Isidore Szczepaniak [5,6] and Marc Webber [5,6]

1 Pacific Mammal Research, Anacortes, WA 98221, USA
2 Cascadia Research Collective, Olympia, WA 98501, USA; Calambokidis@cascadiaresearch.org (J.C.); JHuggins@cascadiaresearch.org (J.L.H.)
3 Department of Biology, Portland State University, Portland, OR 97207, USA; dalin@pdx.edu (D.N.D.); duffieldd@pdx.edu (D.A.D.)
4 Marine Mammal Institute, Oregon State University, Newport, OR 97365, USA; Jim.Rice@oregonstate.edu
5 The Marine Mammal Center, Sausalito, CA 94965, USA; szczepaniaki@tmmc.org (I.S.); webberm@tmmc.org (M.W.)
6 Department of Ornithology and Mammalogy, California Academy of Sciences, San Francisco, CA 94118, USA
* Correspondence: cindy.elliser@pacmam.org

Received: 2 May 2020; Accepted: 23 June 2020; Published: 29 June 2020

Abstract: Harbor porpoise (*Phocoena phocoena*) diets are predominantly comprised of small fish species (<30 cm) and squid. However, predation on larger species (up to 63 cm) occurs, raising the question of increased risk of asphyxiation associated with this behavior. Literature was reviewed and stranding data from 1983 to 2020 from the U.S. West Coast (including California, Oregon and Washington) were searched for cases of prey-related asphyxiation of harbor porpoises and analyzed in relation to age, sex, reproductive status and prey species. Twenty-nine cases were documented. Twenty-seven cases involved large prey; non-native American shad caused the asphyxiation in 87% of the cases where the prey species was identified. The majority (92%) of harbor porpoises were females, and at least 83.3% were pregnant or recently post-partum. Reproductively active females may be more likely to attempt potentially risky behavior in order to compensate for their increased energetic needs. Increasing numbers of non-native American shad may pose a unique danger in this region for harbor porpoises not adapted to deal with the challenges of that prey. This may be a cause for concern, as there is likely an interaction between location, age and reproductive status on the diet composition and foraging strategies of harbor porpoises.

Keywords: harbor porpoise; *Phocoena phocoena*; American shad; salmon; asphyxiation; large prey; foraging ecology; strandings

1. Introduction

Harbor porpoises (*Phocoena phocoena*) are found in nearshore waters along the U.S. West Coast from north of Point Barrow, Alaska (AK), and south to Point Conception, California (CA) [1–3], with very few sightings in offshore waters (beyond the 200 m isobath) [3]. It is important to note that along the U.S. West Coast, the continental shelf is narrow (approximately 32 km) compared to locations in the Atlantic (more than 120 km) [4]. Thus, harbor porpoises on the U.S. West Coast are a nearshore species, unlike some harbor porpoises that are seen in offshore waters in the Atlantic (e.g., off Greenland [5]). Along the U.S. West Coast, harbor porpoises are known to mainly prey on small fish species, such as Pacific herring (*Clupea pallasi*), walleye pollock (*Theragra chalcogramma*), Pacific hake (*Merluccius productus*) northern anchovy (*Engraulis mordax*) and Pacific sandlance (*Ammodytes hexapterus*), that average less than 30 cm in length [6–12], a similar finding to other populations worldwide [13–23].

Harbor porpoises have also been known to ingest large (in mass and/or length, e.g., up to 63 cm) prey items [16,17,21,22,24]. Along the U.S. West Coast, harbor porpoises were observed catching large prey items (consumption was not observed) of native salmonid species, including coho (*Oncorhynchus kisutch*) and pink (*Oncorhynchus gorbuscha*) salmon, and non-native American shad (*Alosa sapidissima*) [25]. A recent study documented tags from juvenile Chinook salmon (*Oncorhynchus tshawytscha*) in the forestomach of a harbor porpoise in southwestern Washington [26]. Previous to these cases, there were rare instances of salmon in the stomachs of harbor porpoises in this region: a salmon sp. in the stomach of a stranded harbor porpoise near the Columbia River [27] and another of a coho salmon in the stomach of a by-caught harbor porpoise in the salmon set-net fishery in Washington State [8]. Salmon have not been considered as part of the normal harbor porpoise diet in this region, but [25] supports the idea presented by [26] that salmon may be a more important component of harbor porpoise diet in the Pacific Northwest than previously thought. Occasionally, attempted ingestion of seemingly atypical prey items and/or prey size that resulted in asphyxiation of the harbor porpoise have been documented along the U.S. West Coast: 55.9 cm grey smooth-hound shark (*Mustelus californicus* [28]), 44.5 cm Pacific herring [29], American shad (37.5 cm, [30] and 34.2 cm [31]) and a species of sculpin [32]. The historical asphyxiation records of female harbor porpoises [30,31] and more recent observations of live female harbor porpoises [25] indicate that American shad may be targeted by some individuals, and there may be a female sex bias to this potentially risky behavior.

Large prey (like salmon and American shad) can increase the risk of asphyxiation in odontocete species due to adaptations to their airway and digestive tracts that removes the risk of water accidentally entering the respiratory system during ingestion of prey [33]. While important, these adaptations also make them more susceptible to esophageal obstruction that can lead to asphyxiation as the larynx, or goosebeak, creates a restriction, limiting the size of items that can be swallowed without risk of them becoming lodged around the larynx (which may become dislocated) [33]. Though not a common occurrence, cause of death by asphyxiation on prey items has been documented in a limited number of cetacean species (harbor porpoises [19,28,30–32,34–37], bottlenose dolphins (*Tursiops* sp.) [38–42], long-finned pilot whales (*Globicephala melas*) [43], Guiana dolphins (*Sotalia guianensis*) [44], and beluga whales (*Delphinapterus leucas*) [45]).

Much of what is known about harbor porpoises in the Pacific Northwest comes from stranded animals [46], as harbor porpoise behavior is still poorly understood and often hard to investigate due to their elusive nature. Stranding records are a good resource for investigating causes of death and their frequency within a population and can provide insights into harbor porpoise foraging behavior. American shad in the Pacific Ocean are mainly found on the continental shelf, with the highest catches (based on data from research cruises, and commercial and recreational landings) between Oregon (OR) and British Columbia (BC), Canada, and in San Francisco Bay [47]. To investigate the occurrence of prey-related asphyxiation in harbor porpoises along the U.S. West Coast and the influence of American shad, we reviewed the available stranding records from researchers and stranding organizations in CA, OR and WA, along with historical published accounts; however, note that this was not an exhaustive search of all historical records. To enhance our knowledge of harbor porpoise foraging behavior we describe these cases, the fish species involved, and the relation to age, sex and reproductive status of the harbor porpoises. We hypothesized that the majority of harbor porpoise asphyxiation cases would involve non-native American shad, and that asphyxiation would occur more with reproductively active females than other age/sex classes.

2. Materials and Methods

Necropsies and records of the stranded harbor porpoises were conducted/recorded by individual researchers and stranding network organizations along the U.S. West Coast (Figure 1): Cascadia

Research Collective (CRC), Portland State University (PSU), Oregon State University (OSU), Washington Department of Fish and Wildlife (WDFW) the California Academy of Sciences (CAS), The Marine Mammal Center (TMMC), California State Parks (CSP), and the Museum of Vertebrate Zoology (MVZ). Although there may be some variation in specifics, the organizations generally adhered to similar stranding and necropsy protocols. Data including the date, location, condition, age class, sex, reproductive status, and evidence of human interaction were collected from all stranding events. Necropsies were conducted on fresh animals using established methods [48]. A full suite of samples from available organs, lymph nodes, lesions and injuries were collected for histology, and additional tissues were frozen for ancillary diagnostics (genetics, contaminants, life history, biotoxins, fungal, viral, bacterial screening). Visual identification of stomach contents was documented for some cases. Internal exams with minimal sampling were conducted on animals that were decomposed and/or heavily scavenged. Evidence for human interaction (vessel interaction, gunshot wounds, fishery interaction wounds related to fishing gear or fishing gear attached to the animal, and other) was noted, whether or not it was the cause of death.

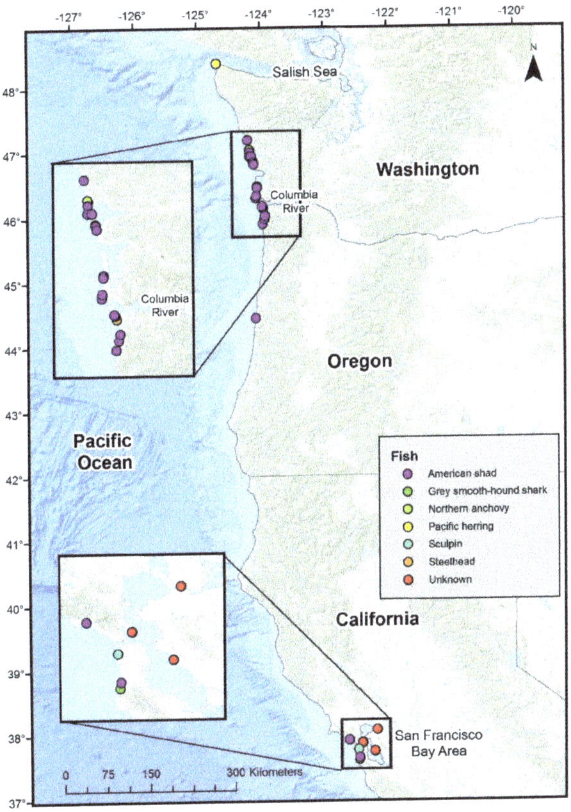

Figure 1. Map of U.S. West Coast with locations (and fish species involved) of asphyxiation cases of harbor porpoises in CA, OR and WA.

3. Results

We documented 24 cases between 1983 and 2020 (numerical case numbers 1–24) from stranding records, in addition to 5 cases previously published (alphabetic case numbers A–E), totaling 29 cases of fatal prey-related asphyxiation in harbor porpoises along the U.S. West Coast since 1937 (Figure 1, Table 1). For the 28 cases where the month and day of stranding are known (all but case B), asphyxiated harbor porpoises were found in every month except November, and 13 (46.4%) of these occurred in spring (March–May). The next most common season was summer (June–August) with eight (28.6%) cases. The number of cases in other seasons were more evenly distributed: three (10.7%) in winter (December–February), and four (14.3%) in fall (September–November). Five cases (10, 15–18) also had evidence of fishery interaction and eight (4, 6, 8, 11, 15, 16, 19, 24) had pathological findings. The contribution of these additional findings to the asphyxiation events are not known. Visual identification of stomach contents for five cases were documented. Case 2 had a pilchard (*Sardinops sagax*) and squid beaks (*Loligo* sp.), case 13 had squid beaks and fresh anchovy, case 18 had an intact American shad, case 21 had fresh and partially digested anchovy and case 24 had partially digested anchovy.

Two types of asphyxiation were found: large fish lodged in the throat, where the goosebeak is often dislocated (cases A–D, 1–20, 22–24), and smaller fish moving up to obstruct the nasal passage (cases E, 21). Twenty-seven cases (93.1%) resulted from attempted ingestion of large prey items and subsequent asphyxiation (cases A–D, 1–20, 22–24); 25 were confirmed as cause of death and 2 where it was the likely cause of death, but the carcass was not collected (cases 5, 9, Table 1). Many of these incidences occurred in the last four years (n = 9, 33%). In all cases the large fish was lodged in the harbor porpoise's throat, sometimes with the tail still visibly sticking out of the mouth (Figures 2, 3, 5 and 6). The fish was known to be oriented head first in all but two cases (92.6%, in cases D and 9 the orientation was not documented). In 18 (66.7%) cases the larynx or goosebeak was dislocated (Figure 2), sometimes with the fish extending past the proximal end of the lungs. Dry drowning was suspected in the four cases (13, 21, 22, 24) for which the condition of the lungs was noted. Observations included very aerated lung tissue, white stable froth in main airways and trachea and no fluid, with the exception of one case (22) that had a small amount of fluid at the base of one lung.

American shad was the species causing the asphyxiation in 87% (n = 20, cases B–C, 1–4, 6, 8, 10–15, 17–19, 22–24) of the large prey ingestion cases where the prey species was identified. The American shad ranged in length from 33 to 49 cm (n = 18 cases where the whole fish was measured) and weighed from 363 to 1269 g (n = 10 cases where the whole fish was weighed). The other identified large fish species was a steelhead (*Oncorhynchus mykiss*; 70 cm, 3200 g, case 16), grey smooth-hound shark (55.9 cm, case A), Pacific herring (44.5 cm, case D) and four unknown species (cases 5, 7, 9, 20). For the cases involving large prey items and known sex of the harbor porpoise (n = 25), 23 were female (92%, cases A–D, 1–4, 6, 8, 11–20, 22–24) and 2 were male (8%, cases 7, 10). Size of the female harbor porpoises ranged from 101.6 to 181 cm, and males (both subadult) were 121.7 and 130 cm in size (Table 1). Reproductive status was determined for 18 females that attempted to ingest the large prey items: 11.1% (n = 2) were not pregnant (cases C, 14) and 66.7% (n = 12) were confirmed to be pregnant (cases D, 4, 6, 8, 11–13, 15–16, 18–19, 24). Four other cases were likely recently pregnant: one with endometrial hemosiderosis suggestive of recent prior pregnancy (case 3), one with a distended uterus (case 17), one possibly post-partum (case 22) and one with a recent corpus luteum (case 23) (Table 1). Combining the 12 pregnancies with three likely post-partum females raises the percentage of cases of reproductively active females to 83.3% (n = 15), or 88.9% (n = 16) if the female with a recent corpus luteum is included.

Table 1. Cases of prey-related asphyxiation of harbor porpoises including life history characteristics, fish species (size and weight when available) and details of the necropsy findings along the U.S. West Coast. * Cases A–E are previously published accounts [28–32]. For large prey items, the fish was stuck head first in all but two cases (D and 9), where the orientation was not documented.

Case	Date (Year/Month/Day)	State	Contributor	HP Sex	HP Age	Length (cm)	Carcass Condition	♀Reproductive Status	Fish Species	Fish Length (cm)	Fish Weight (g)	Details
A	1937/3/9	CA	28 *	F	SubA	101.6	Fresh	Unknown	Grey smooth-hound shark	55.9	NA	Tail of the shark was sticking out of mouth, head was partly in stomach. Froth in lung indicated suffocation.
B	1948	WA	30 *	F	A	Unknown	Unknown	Unknown	American shad	37.5	NA	American shad stuck head first in throat.
C	1952/6/8	WA	31 *	F	A	147	Fresh	Not visibly pregnant	American shad	34.2	680	American shad stuck head first in throat.
D	1979/12/19	WA	29 *	F	A	137	Fresh	Pregnant	Pacific herring	44.5	NA	Fish firmly stuck in throat, pregnant with 20.3 cm well developed fetus.
E	2008/6/30	CA	32 *	M	SubA	132	Fresh	NA	*Sculpin* sp.	~13	NA	Sculpin lodged in nasal passages, completely obstructing it.
1	1983/5/9	WA	WDFW	F	A	150	Moderate	Unknown	American shad	NA	NA	American shad stuck in throat with tail sticking out of the mouth.
2	1992/9/5	CA	CAS/MVZ	F	A	153	Moderate	Unknown	American shad	NA	NA	American shad stuck in throat with tail sticking out of the mouth; A pilchard and several squid beaks in stomach.
3	2005/6/19	CA	TMMC	F	A	165	Fair	Recently pregnant, endometrial hemosiderosis	American shad	41	NA	American shad wedged in esophagus, tail was sticking out of the mouth, goosebeak was dislocated.
4	2006/1/12	WA	CRC	F	A	161	Fresh	Pregnant	American shad	NA	NA	American shad stuck in throat head first, goosebeak dislocated.
5	2007/5/5	WA	CRC	U	A	150 (est.)	Fresh	Unknown	Unknown large fish	NA	NA	Large fish was lodged in mouth, only the tail was visible. Reported by reliable source, carcass not collected.
6	2008/5/21	OR	PSU	F	A	162.5	Moderate	Pregnant	American shad	47	943	American shad stuck head first in throat, goosebeak dislocated.
7	2009/3/6	CA	CAS	M	SubA	130	Moderate	NA	Unknown large fish	35	NA	Fish lodged in esophagus with tail extending into the oral cavity, goosebeak dislocated and esophagus was distended.
8	2009/3/14	OR	PSU	F	A	151.4	Moderate	Pregnant	American shad	43	852	American shad stuck head first in throat, goosebeak dislocated.
9	2010/6/1	CA	CSP	U	SubA	122	Fresh	Unknown	Unknown large fish	NA	NA	Fish stuck in throat, carcass not collected.
10	2011/9/13	WA	PSU	M	SubA	121.7	Moderate	NA	American shad	33	398	American shad stuck head first in throat, goosebeak dislocated, fishery interaction.
11	2013/4/29	OR	OSU	F	A	163	Fresh	Pregnant	American shad	38	NA	American shad stuck in throat head first obstructing the trachea and bruising the goosebeak.
12	2014/2/1	WA	PSU	F	A	162.5	Moderate	Pregnant	American shad	42.5	755	American shad stuck head first in throat, goosebeak dislocated.
13	2014/3/20	WA	CRC	F	A	162	Fresh	Pregnant	American shad	35.5	NA	American shad stuck head first in throat, goosebeak dislocated. Head of fish was down level with the lungs, some regurgitated food in the esophagus. Squid beaks and fresh anchovy in stomach. Figure 2.
14	2014/5/9	WA	CRC	F	A	181	Moderate/ Advanced	Not pregnant	American shad	40 *	NA	American shad stuck head first in throat, goosebeak dislocated. Minimal exam due to carcass condition. * scavenged, estimated length. Figure 3.

Table 1. *Cont.*

Case	Date (Year/Month/Day)	State	Contributor	HP Sex	HP Age	Length (cm)	Carcass Condition	♀Reproductive Status	Fish Species	Fish Length (cm)	Fish Weight (g)	Details
15	2016/5/23	OR	PSU	F	A	168	Moderate	Pregnant	American shad	46.7	870	American shad stuck head first in throat, goosebeak dislocated. Fishery interaction.
16	2016/7/3	OR	PSU	F	A	163	Fresh	Pregnant	Steelhead	70	3200	Steelhead stuck head first in throat, goosebeak dislocated. Fishery interaction. A full-term fetus was pressing on the caudal end of the forestomach, preventing full expansion of the stomach by 2.5 to 5 cm and thereby preventing the fish from being completely swallowed.
17	2016/7/16	OR	PSU	F	A	169.2	Advanced	Recently pregnant, distended uterus	American shad	49	837	American shad stuck head first in throat, goosebeak dislocated. Fishery interaction. Recently pregnant with distended uterus.
18	2017/4/28	WA	PSU	F	A	174.5	Moderate	Pregnant	American shad	30.9	870	3/4 of American shad stuck head first in throat, goosebeak dislocated. Fishery interaction; 649 g intact American shad also in forestomach.
19	2018/5/31	WA	PSU	F	A	166	Fresh	Pregnant	American shad	41.1	1269	American shad stuck head first in throat, goosebeak dislocated.
20	2018/8/15	CA	CAS	F	A	152.5	Advanced	Unknown	Unknown large fish	~30	NA	Fish stuck head first in throat, tail sticking out of the mouth, goosebeak dislocated.
21	2018/10/3	WA	CRC	F	A	184	Fresh	Unable to determine	Northern anchovy	7.6	NA	Fish lodged in blowhole. It was suspected that the harbor porpoise was feeding on the anchovy when attacked. The goosebeak was dislocated and there was bruising and edema around the blowhole at the top of the head and bruising in the back of the throat (Figure 4). There were pre-mortem bite marks from a broadnose seven gill shark. Reproductive status of the harbor porpoises was unknown because the shark had bitten out that area of the abdomen. Fresh and partially digested anchovy in stomach.
22	2019/6/28	WA	CRC	F	A	174.5	Fresh	Possibly post-partum	American shad	49	NA	American shad stuck head first in throat, goosebeak dislocated and the fish was down the esophagus past the heart (Figure 5). Shad was a gravid female.
23	2019/9/19	OR	PSU	F	A	167.5	Moderate	Not pregnant, recent corpus luteum	American shad	46.5	686	American shad stuck head first in throat, goosebeak dislocated. Not pregnant or lactating, but had recent corpus luteum.
24	2020/4/25	WA	CRC	F	A	165	Fresh	Pregnant	American shad	49	NA	American shad (gravid female) stuck head first in throat, goosebeak dislocated (Figure 6). Stomach full of partially digested anchovy. Third-trimester fetus.

The last two asphyxiation cases resulted from a small fish species blocking the nasal passage. One case was the presumptive regurgitation, or pushed by force up the esophagus, of a northern anchovy (*Engraulis mordax*; 7.6 cm, weight unknown, Figure 4), resulting from a shark attack from a probable broadnose seven gill shark (*Notorynchus cepedianus*). Determination of the shark species involved was based on the size and shape of bites, characteristics of the tooth marks on the skin, blubber and internal organs and the geographic location of the incident (case 21). The second was a species of sculpin (~13 cm, weight unknown) that became lodged in the nasal passage (case E).

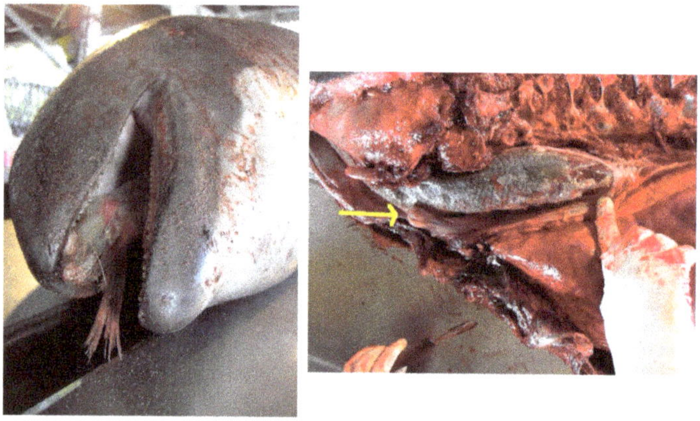

Figure 2. Case 13: 20 March 2014, pregnant adult female harbor porpoise from Ocean Shores, WA. American shad (35.5 cm) dislocated the goosebeak (arrow) from the blowhole area. Photo credit CRC.

Figure 3. Case 14: 9 May 2014, American shad (partially eaten by scavengers post-mortem, estimated 40 cm) that was found in an adult female harbor porpoise's throat causing asphyxiation (Westport, WA). Photo credit CRC.

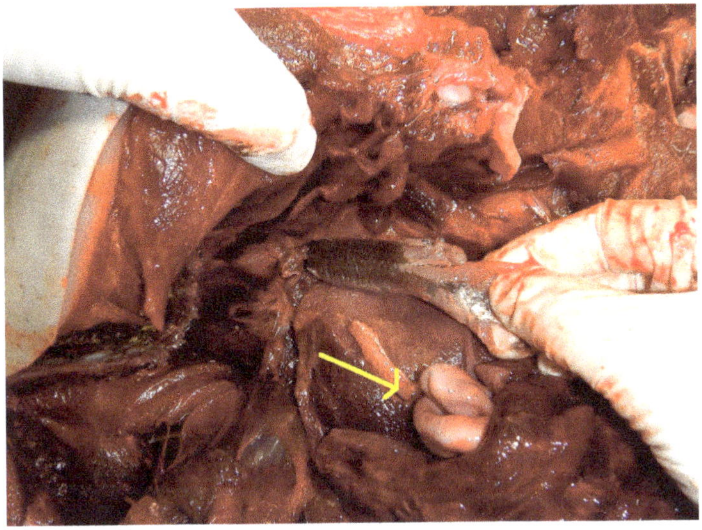

Figure 4. Case 21: October 3, 2018 adult female harbor porpoise from Ocean City, WA, asphyxiated on northern anchovy (7.6 cm) lodged in blowhole, goosebeak (arrow) dislocated with bruising and edema around the blowhole at the top of the head and bruising in the back of the throat. Photo credit CRC.

Figure 5. Case 22: 28 July 2019, possibly post-partum adult female harbor porpoise from Ocean Shores, WA, with 49 cm American shad stuck in in the throat (partially eaten by scavengers post-mortem), dislocating the goosebeak and stuck down the esophagus and past the heart. Photo credit CRC.

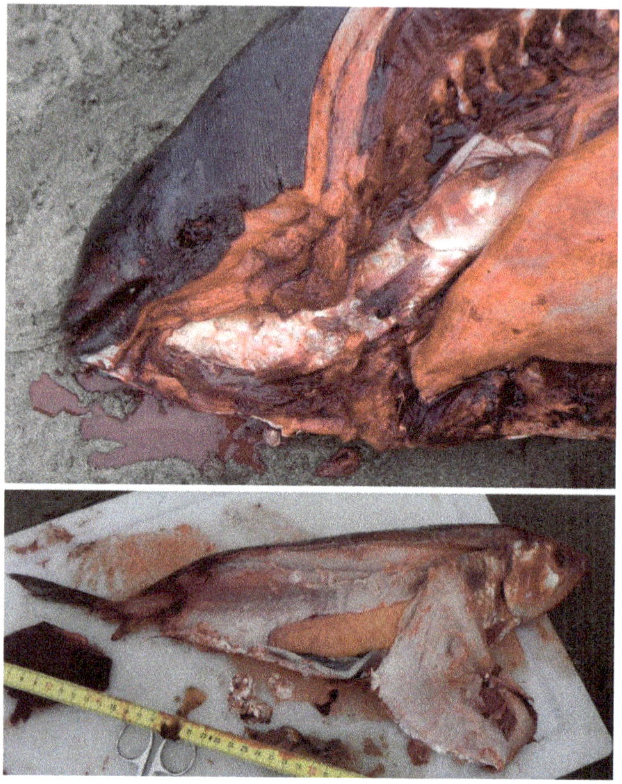

Figure 6. Case 24: 25 April 2020, third trimester pregnant adult female harbor porpoise from Ocean Shores, WA, with 49 cm American shad gravid female stuck in in the throat, dislocating the goosebeak.

4. Discussion

This study represents the largest set of records (29 total) published to date of prey-related asphyxiation in harbor porpoises. Of the large fish that were identified, 87% were American shad, showing a marked bias in the prey species responsible. These data support the idea that the events observed by [25] of harbor porpoises catching American shad and salmonid species along the U.S. West Coast were indeed foraging behavior, and that these species should be considered as potential prey species of harbor porpoise in this region. Further, our results suggest an age/sex bias to this behavior as 92% of harbor porpoises were female (all adult, except one), and that at least 83.3% of these females were in some stage of reproductive activity (pregnant or recently pregnant/post-partum). This bias towards females (and particularly pregnant females) is not reflected in overall stranding records for this region. In CA and the outer coast of WA, a total of 405 harbor porpoises were examined between 2005 and April 2020: 49.1% male, 45.2% female and 5.7% unknown sex and in WA, 45.7% of the adult females were pregnant (unpublished data). Although the cases of asphyxiation reported here represent a small portion of the total number of examined harbor porpoises, due to variability in carcass and environmental conditions, cetaceans that float or refloat and are cast ashore are likely a small subset of the actual mortality [49,50] and additional instances may have occurred and have gone undetected. Thus, our results may underrepresent this source of mortality.

Harbor porpoises are the only cetaceans in the region with documented American shad asphyxiation, which may be the result of multiple factors. Since the introduction of American shad to the Pacific in the Sacramento River in 1871 (and now are found from Baja California, Mexico to Siberia) [51], these fish have overlapped in range with harbor porpoises as both are mainly found on the continental shelf, and consequently are more likely to interact with harbor porpoises than with cetaceans found in deeper water. The harbor porpoise preference for nearshore habitat also means that dead specimens of this species wash ashore more frequently, and in better post-mortem examination condition than offshore odontocetes [50]. Finally, as the world's second smallest cetacean, harbor porpoises may be ill-equipped to ingest American shad which averages 50–61.7 cm and up to 5500 g [52] simply due to predator–prey body size ratios and/or the morphology of American shad compared to other large prey items.

Remarkably, in nearly 40 years of diet studies, there is no evidence of American shad as prey for harbor porpoises along the U.S. West Coast based on otolith remains [6–9,11], stable isotope ratios [10] or PCR-based molecular techniques [12]. The rate at which otoliths erode varies by prey species [11] and American shad otoliths are relatively small, so this species may be more easily missed, or not recognized, in this type of diet study. Alternatively, the otoliths may not be present due to beheading the fish prior to consumption. Beheading and not ingesting the head of the fish has been suggested for harbor porpoises [13], and documented for rough-toothed dolphins (*Steno bredanensis*) [53], Amazon river dolphins (*Inia geoffrensis*) [54] and bottlenose dolphins [55]. However, this behavior has not been documented with American shad and the small spade-shaped teeth of harbor porpoises would make it difficult to perform the ripping or tearing action necessary to behead a large fish. In every case in the current study where the orientation of the fish was noted, the fish was head first down the harbor porpoise's throat. With the head hitting the gastric juices first, the head and the small otoliths may degrade quickly and thus not be present, or identifiable, when examined (a more likely scenario than beheading as suggested by [13]). Nevertheless, diet studies that do not rely on otolith remains have not found American shad in the diet of harbor porpoises in CA [10,12]. Alternatively, the lack of evidence of this prey species may be due to this being a more recent phenomenon, opportunistic captures, and/or that there are only certain groups or subpopulations (like reproductively active adult females in this study) that utilize this strategy (e.g., [55]).

In our study, 85.2% of the cases of asphyxiation on large prey items involved adult harbor porpoises. Variation in diet between adult and juvenile harbor porpoises has been reported [21,24]. A positive correlation between the length of harbor porpoises and some of their prey [21], along with the fact that adults are presumably more experienced foragers, make it more likely that adults would target larger prey than juveniles. However, if size was the only determinant (even with harbor porpoise females being marginally larger than males [56]), we would not see the striking bias towards females (92%) that we documented. Other studies have shown a similar bias in sex and maturity influencing variation in harbor porpoise diet. In Icelandic waters, adult females had more diverse diets than males or subadult females, and two females (one pregnant) had ingested the largest prey items [21]. Other cases of asphyxiation are bias towards females (in the North Sea for common sole, *Solea solea* and cod, *Gadus morhua* (reproductive state not documented) [36]; in the Western Baltic, a lactating female likely asphyxiated (full necropsy was not completed) on a 59 cm European eel [37]). Together, these data suggest that adult female harbor porpoises may be more likely to target larger prey species and that this may be even further biased by reproductive state.

Over 83% of females of known reproductive state were either pregnant or recently pregnant/post-partum. With yearly or biyearly birth cycles [57,58], harbor porpoises spend much of their adult lives both pregnant and lactating [59,60] and thus are likely to be in some stage of reproductive activity at any given time of year. Bioenergetic models (incorporating basal metabolism, thermoregulation and reproduction) comparing all adult reproductive states reveal that male harbor porpoises have the lowest total energetic costs while females that were simultaneously lactating and pregnant had the highest [61]. The energetic cost of lactation exceeds the costs of pregnancy [61] and lactation and post-partum pregnancies require increased food intake [61,62]. In bioenergetic models, pregnant and lactating females had the highest biomass intake per day needed (4.1 ± 1.6 kg, feeding on energy rich species like northern anchovy and rockfish) to travel 68.45 ± 26.45 km/day [61].

To obtain those 4.1 kg, a pregnant and lactating harbor porpoise would need to expend energy to capture and consume approximately 456 northern anchovies (average 9 g each), or one American shad (up to 5.5 kg). In contrast, a male (needing 1.7 ± 0.6 kg per day) would only have to capture and consume approximately 189 northern anchovies, while an American shad could be as much as 3-fold the amount of mass needed daily. Note that these estimates are based on mass alone, and do not take into consideration the differences in energy density of the prey species, which may vary. Regardless, it is clear that reproductively active females need to take in a greater mass of high quality fish than do other age/sex classes.

Reproductively active female harbor porpoises may have altered foraging behaviors compared to other sex/age classes in order to meet these increased energetic needs. They may increase foraging time, consume more prey items, ingest different prey species not seen in other harbor porpoises, and/or select prey of much larger than average length and mass [14,62–64]. Females with calves have been shown to utilize habitats with reduced water movement [25,63] and have shown learned strategies to reduce unproductive foraging time [62,63]. These behaviors may impact what prey (and quantity) they take, along with how productive, in terms of energy balance, the foraging will be. This may be particularly important for pregnant females because those in poor body condition may reduce their energetic investment in the fetus, presumably to increase their own survivability [65]. Thus, pregnant females (which constituted 66.7% of females of known reproductive status catching large fish in the current study) may be more likely to attempt a potentially risky behavior (such as ingesting American shad) in order to gain the nutritional payoff that will help them maintain their health, and the health of the unborn calf. Many predators of different age, sex and reproductive classes will vary in their willingness to attack dangerous prey [66]. Our results are thus not surprising, as the constraints inherent in the lives of reproductively active female harbor porpoises make it more likely, if not necessary, for them to attempt capturing larger prey than other age/sex classes.

However, the relative risk is unclear, as there is evidence that harbor porpoises can successfully ingest larger prey items [8,16,17,21,22,24,25]. This has also been observed in asphyxiation events, as species that caused the death of the harbor porpoises have been found partially digested in the stomach (Table 1 cases 16 and 18; [36]), indicating that the species (and size) can sometimes be consumed without issue. These examples indicate that harbor porpoises are likely able to successfully take these larger prey, but may not be reflected in stranding data that tend to be biased toward compromised individuals, or in behavioral studies of harbor porpoises due to the difficulties in observing consumption of fish [25]. This could be another reason why our results may underrepresent the prevalence of this behavior.

Attempted ingestion of American shad appears to be an increasing phenomenon, as there were just four cases documented prior to 2005 and a subsequent increase after that, with 33% occurring between 2016 and 2020. This correlates with the continued increase in American shad, which have become the single largest spawning run of any local anadromous fish (over 4 million spawning adults passing Bonneville Dam in the Columbia River from 2003 to 2006), often dwarfing those of the salmon runs [51]. Increases in the population density of a particular species, or a steep decline in other, regular prey populations, may influence the prey choice of harbor porpoises. These factors have been suggested as reasons for asphyxiations in harbor porpoises [36] and long-finned pilot whales [43]. Alternatively, it could be related to increased inter- or intraspecific competition, as has been suggested for bottlenose dolphins engaging in riskier foraging behavior [55,67]. However, this could be an artifact of increased harbor porpoise presence, as it is difficult to tease apart true increases in a certain event from increased interest and observations. For example, an unusual mortality event along the Pacific Northwest in the period 2006–2007 was determined not to be an actual unusual mortality event, but rather a result of an increasing harbor porpoise population, expansion of harbor porpoise range and a more well-established stranding network [46]. The increasing frequency of the asphyxiation events documented in this study may indicate either shifts in native prey availability and/or in prey choice by a particular class of individuals (in this case reproductively active females), but more research is needed to understand the driving force for these behaviors.

5. Conclusions

Our review of stranding records and necropsy findings on the U.S. West Coast revealed that American shad was responsible for the majority of harbor porpoise asphyxiations. This prey is an introduced species in this region and may pose a unique danger for harbor porpoises not adapted to deal with the challenges of this prey. The majority of cases involved pregnant or recently post-partum adult female harbor porpoises and this may reflect an attempt to consume large prey during periods of greater energetic need. The incidence of these events has increased, with 33% occurring between 2016 and 2020, and likely reflects increasing abundance of this introduced species. It is striking that American shad has not been documented as harbor porpoise prey (and to our knowledge no asphyxiation events have been published) in the Atlantic, where the fish is native. Updated large-sample diet studies are needed for both coasts to help determine the prevalence of American shad (or other large prey items, like salmon) in the harbor porpoise diet. We recommend that future research should focus on both harbor porpoise and American shad morphology, energetics and behavior, along with comparisons of native and non-native prey abundance/distribution patterns and marine mammal stranding records from the U.S. West and East Coasts. This is critical to gain a better understanding of harbor porpoise foraging ecology and will help to elucidate the reasons for this behavior and the relative risk it carries for harbor porpoise populations.

Author Contributions: Conceptualization, C.R.E., D.N.D., D.A.D., J.L.H., J.R., I.S. and M.W.; project administration, C.R.E.; resources, D.N.D., D.A.D., J.L.H., J.R., I.S. and M.W.; supervision, C.R.E.; validation, C.R.E., J.C., D.N.D., D.A.D., J.L.H., J.R., I.S. and M.W.; visualization, C.R.E., J.C., D.N.D., J.L.H., J.R., I.S. and M.W.; writing—original draft, C.R.E.; writing—review and editing, C.R.E., J.C., D.N.D., D.A.D., J.L.H., J.R., I.S. and M.W. All authors have read and agreed to the published version of the manuscript.

Funding: Stranding response activities since 2004 were supported by several grants from the John H. Prescott Marine Mammal Rescue Assistance Grant Program (NOAA fisheries).

Acknowledgments: Thanks to all the staff, volunteers, and supporters of Pacific Mammal Research, especially S. Elliser and Marathon Petroleum Foundation. Thanks to all the staff and volunteers from Cascadia Research for assistance with response and examinations, specifically E. Keene. Thanks to former and current Seaside Aquarium staff for assisting Portland State University with stranding response, specifically K. Chandler, J. Hussa and T. Boothe. Thanks to all the staff and volunteers from The Marine Mammal Center and The California Academy of Sciences. Thanks to the efforts of the many volunteers of the Oregon Marine Mammal Stranding Network. Thanks to S. Sveegaard for the personal communication, W. Keener and C. Gallagher for reviews of the manuscript, D. Anderson for assistance in creating the map, D. Lowry from WDFW for shark and fish identification, M. Myrsell from the Westport Aquarium for assistance with carcass recovery and examination of one case and D. Lambourn from Washington Department of Fish and Wildlife for details on the case from their historical records that all improved the manuscript. We appreciate the suggestions of two anonymous reviewers that improved the manuscript.

Conflicts of Interest: The authors declare no conflict of interest.

References

1. Osmek, S.; Calambokidis, J.; Laake, J.; Gearin, P.; Delong, R.; Scordino, J.; Jeffries, S.; Brown, R. *Assessment of the Status of Harbor Porpoise (Phocoena phocoena) in Oregon and Washington Waters*; NOAA Technical Memorandum NMFS-AFSC-76; U.S. Department of Commerce: Washington, DC, USA, 1996.
2. Barlow, J. Harbor porpoise, *Phocoena phocoena*, abundance estimation for California, Oregon, and Washington I. Ship surveys. *Fish. Bull.* **1988**, *83*, 417–432.
3. Forney, K.A.; Carretta, J.V.; Benson, S.R. *Preliminary Estimates of Harbor Porpoise Abundance in Pacific Coast Waters of California, Oregon, and Washington, 2007–2012*; NOAA Technical Memorandum NMFS-SWFSC-537; U.S. Department of Commerce: Washington, DC, USA, 2014.
4. Encyclopedia Britannica. Available online: https://www.britannica.com/science/continental-shelf (accessed on 1 June 2020).
5. Nielsen, N.H.; Teilmann, J.; Sveegaard, S.; Hansen, R.G.; Sinding, M.H.S.; Dietz, R.; Heide-Jørgensen, M.P. Oceanic movements, site fidelity and deep diving in harbour porpoises from Greenland show limited similarities to animals from the North Sea. *Mar Ecol. Prog. Ser.* **2018**, *597*, 259–272. [CrossRef]

6. Jones, R.E. Food habits of smaller marine mammals from northern California. *Proc. Calif. Acad. Sci.* **1981**, *42*, 409–433.
7. Dorfman, E.J. Distribution, Behavior, and Food Habits of Harbor Porpoises (*Phocoena phocoena*) in Monterey Bay. Master's Thesis, San Jose State University, San Jose, CA, USA, 1990.
8. Gearin, P.J.; Melin, S.R.; Delong, R.L.; Kajimura, H.; Johnson, M.A. Harbor porpoise interactions with a chinook salmon set-net fishery in Washington State. *Rep. Int. Whal. Commn. Special* **1994**, 427–438.
9. Walker, W.A.; Hanson, M.B.; Baird, R.W.; Guenther, T.J. Food habits of the harbor porpoise, *Phocoena phocoena*, and Dall's porpoise, *Phocoenoides dalli*, in the inland waters of British Columbia and Washington. *Alask. Fish. Sci. Center Process. Rep.* **1998**, *98-10*, 63–75.
10. Toperoff, A.K. Examination of Diet of Harbor Porpoise (*Phocoena phocoena*) from Central California Using Stomach Content and Stable Isotope Analysis from Multiple Tissues. Master's Thesis, San Jose State University, San Jose, CA, USA, 2002.
11. Nichol, L.M.; Hall, A.M.; Ellis, G.M.; Stredulinsky, E.; Boogaards, M.; Ford, J.K. Dietary overlap and niche partitioning of sympatric harbour porpoises and Dall's porpoises in the Salish Sea. *Prog. Oceanogr.* **2013**, *115*, 202–210. [CrossRef]
12. Oliaro, J.F. Identification of Harbor Porpoise (*Phocoena phocoena*) Prey Species Using PCR Based Molecular Technique. Master's Thesis, San Francisco University, San Francisco, CA, USA, 2013.
13. Smith, G.J.D.; Gaskin, D.E. The diet of harbor porpoises (*Phocoena phocoena* (L.)) in coastal water of Eastern Canada, with special reference to the Bay of Fundy. *Can. J. Zool.* **1974**, *52*, 777–782. [CrossRef]
14. Recchia, C.A.; Read, A.J. Stomach contents of harbour porpoises, *Phocoena phocoena* (L.), from the Bay of Fundy. *Can. J. Zool.* **1989**, *67*, 2140–2146. [CrossRef]
15. Smith, R.J.; Read, A.J. Consumption of euphausiids by harbour porpoise (*Phocoena phocoena*) calves in the Bay of Fundy. *Can. J. Zool.* **1992**, *70*, 1629–1632. [CrossRef]
16. Fontaine, P.M.; Hammill, M.O.; Barrette, C.; Kingsley, M.C. Summer diet of the harbour porpoise (*Phocoena phocoena*) in the estuary and the northern Gulf of St. Lawrence. *Can. J. Fish.* **1994**, *51*, 172–178. [CrossRef]
17. Aarefjord, H.; Bjørge, A.J.; Kinze, C.C.; Lindstedt, I. Diet of the harbour porpoise (*Phocoena phocoena*) in Scandinavian waters. *Oceanogr. Lit. Rev.* **1996**, *10*, 1041.
18. Martin, A.R. *The Diet of Harbour Porpoises (Phocoena phocoena) in British Waters*; SC/47/SM48; International Whaling Commission: Cambridge, UK, 1996.
19. Benke, H.; Siebert, U.; Lick, R.; Bandomir, B.; Weiss, R. The current status of harbor porpoises (*Phocoena phocoena*) in German waters. *Arch. Fish. Mar. Res./Arch. Fisch. Meeresforsch.* **1998**, *46*, 97–123.
20. Gannon, D.P.; Craddock, J.E.; Read, A.J. Autumn food habits of harbor porpoises, *Phocoena phocoena*, in the Gulf of Maine. *Fish. Bull.* **1998**, *96*, 428–437.
21. Víkingsson, G.A.; Ólafsdóttir, D.; Sigurjónsson, J. Geographical, and seasonal variation in the diet of harbour porpoises (*Phocoena phocoena*) in Icelandic coastal waters. *NAMMCO Sci. Publ.* **2003**, *5*, 243–270. [CrossRef]
22. Sveegaard, S.; Andreasen, H.; Mouritsen, K.N.; Jeppesen, J.P.; Teilmann, J.; Kinze, C.C. Correlation between the seasonal distribution of harbour porpoises and their prey in the Sound, Baltic Sea. *Mar. Biol.* **2012**, *159*, 1029–1037. [CrossRef]
23. Wisniewska, D.M.; Johnson, M.; Teilmann, J.; Rojano-Doñate, L.; Shearer, J.; Sveegaard, S.; Madsen, P.T. Ultra-high foraging rates of harbor porpoises make them vulnerable to anthropogenic disturbance. *Curr. Biol.* **2016**, *26*, 1441–1446. [CrossRef] [PubMed]
24. Andreasen, H.; Ross, S.D.; Siebert, U.; Andersen, N.G.; Ronnenberg, K.; Gilles, A. Diet composition and food consumption rate of harbor porpoises (*Phocoena phocoena*) in the western Baltic Sea. *Mar. Mammal Sci.* **2017**, *33*, 1053–1079. [CrossRef]
25. Elliser, C.R.; Hessing, S.; MacIver, K.H.; Webber, M.A.; Keener, W. Harbor porpoises (*Phocoena phocoena vomerina*) catching and handling large fish on the U.S. West Coast. *Aquat. Mamm.* **2020**, *46*, 191–199. [CrossRef]
26. D'Alessandro, D.N.; Duffield, D.A. Salmonid passive integrated transponder tags and coded wire tags found in the forestomach of a harbor porpoise (*Phocoena phocoena*) in southwestern Washington. *Fish. Bull.* **2019**, *117*, 303–307. [CrossRef]

27. Treacy, S.D. Feeding habits of marine mammals from Grays Harbor, Washington to Netarts Bay Oregon. Marine mammals and their interactions with fisheries of the Columbia River and adjacent waters, 1980–1982. NOAA: Northwest Alaska Fish. Sci. Cent. NWAFC Processed Report 85-04. *Natl. Mar. Fish. Serv.* **1985**, 149–198. Available online: https://archive.fisheries.noaa.gov/afsc/Publications/ProcRpt/NWAFCPR85-04.pdf (accessed on 2 May 2020).
28. Orr, R.T. A porpoise chokes on a shark. *J. Mammal.* **1937**, *18*, 370. [CrossRef]
29. Hult, R.W.; Dupey, S.E.; Badley, R.W. Mortalities associated with prey ingestion by small cetaceans. *Cetology* **1980**, *38*, 1–2.
30. Scheffer, V.B.; Slipp, J.W. The whales and dolphins of Washington State with a key to the cetaceans of the West Coast of North America. *Am. Midl. Nat.* **1948**, *39*, 257–337. [CrossRef]
31. Scheffer, V.B. Measurements and stomach contents of eleven delphinids from Northeast Pacific. *Murrelet* **1953**, *34*, 27–30. [CrossRef]
32. Wilkin, S.M.; Cordaro, J.; Gulland, F.M.D.; Wheeler, E.; Dunkin, R.; Sigler, T.; Casper, D.; Berman, M.; Flannery, M.; Fire, S.; et al. An unusual mortality event of harbor porpoises (*Phocoena phocoena*) off central California: Increase in blunt trauma rather than an epizootic. *Aquat. Mamm.* **2012**, *38*, 301–310. [CrossRef]
33. MacLeod, C.D.; Reidenberg, J.S.; Weller, M.; Santos, M.B.; Herman, J.; Goold, J.; Pierce, G.J. Breaking symmetry: The marine environment, prey size, and the evolution of asymmetry in cetacean skulls. *Anat. Rec.* **2007**, *290*, 539–545. [CrossRef] [PubMed]
34. Siebert, U.; Wünschmann, A.; Weiss, R.; Frank, H.; Benke, H.; Frese, K. Post-mortem findings in harbour porpoises (*Phocoena phocoena*) from the German North and Baltic Seas. *J. Comp. Path.* **2001**, *124*, 102–114. [CrossRef] [PubMed]
35. Ryan, C.; Bolin, V. Stranded harbour porpoise (*Phocoena phocoena (Grey)*) with throat obstructed by a scad (*Trachurus trachurus* (L.)). *Ir. Nat. J.* **2014**, *33*, 155.
36. Roller, M.; Gross, S.; Reckendorf, A.; Andreasen, H.; Wohlsein, P.; Siebert, U. Fatal asphyxiation with laryngeal displacement caused by platvissen in harbour porpoises. In Proceedings of the 31st Annual Conference of the European Cetacean Society, Middelfart, Denmark, 1–3 May 2017.
37. Sveegaard, S. *Personal Communication*; Aarhus University: Aarhus, Denmark, 2020.
38. Byard, R.W.; Gilbert, J.D.; Gibbs, S.E.; Kemper, C.M. Cetacean Café Coronary. *J. Clin. Forensic Med.* **2003**, *10*, 85–88. [CrossRef]
39. Byard, R.W.; Tomo, I.; Kemper, C.M.; Gibbs, S.E.; Bossley, M.; Machado, A.; Hill, M. Unusual causes of fatal upper aerodigestive tract obstruction in wild bottlenose dolphins (*Tursiops aduncus*). *Forensic Sci. Med. Pathol.* **2010**, *6*, 207–210. [CrossRef]
40. Watson, A.; Gee, L.E. Laryngeal displacement and asphyxiation by a beheaded sheepshead (*Archosargus probatocephalus*) in a bottlenose dolphin (*Tursiops truncatus*). *Aquat. Mamm.* **2005**, *31*, 447–452. [CrossRef]
41. Mignucci-Giannoni, A.A.; Rosario-Delestre, R.J.; Alsina-Guerrero, M.M.; Falcon-Matos, L.; Guzmán-Ramírez, L.; Williams, E.H.; Bossart, G.D.; Reidenberg, J. Asphyxiation in a bottlenose dolphin (*Tursiops truncatus*) from Puerto Rico due to choking on a black margate (*Anisotremus surinamensis*). *Aquat. Mamm.* **2009**, *35*, 48–54. [CrossRef]
42. Stolen, M.; St. Leger, S.; Noke Durden, W.; Mazza, T.; Nilson, E. Fatal asphyxiation in bottlenose dolphins (*Tursiops trunactus*) from the Indian River Lagoon. *PLoS ONE* **2013**, *8*, e66828. [CrossRef] [PubMed]
43. IJsseldijk, L.L.; Leopold, M.F.; Bravo Rebolledo, E.L.; Deaville, R.; Haelters, J.; IJzer, J.; Jepson, P.D.; Gröne, A. Fatal asphyxiation in two long-finned pilot whales (*Globicephala melas*) caused by common soles (*Solea solea*). *PLoS ONE* **2015**, *10*, e0141951. [CrossRef] [PubMed]
44. Mariani, D.B.; Guimarães, J.P.; Batista, R.G.; Brum, A.; Groch, K.R.; Díaz-Delgado, J.; Parente, J.E.V. Fatal asphyxia due to laryngeal displacement by large-sized prey in a Guiana dolphin (*Sotalia guianensis*), Brazil. *Cienc. Rural* **2020**, *50*, e20190068. [CrossRef]
45. Rouse, N.; Burek-Huntington, K.A.; Shelden, K.E.W. Asphyxiation of an endangered Cook Inlet beluga whale, *Delphinapterus leucas*. *Mar. Fish. Rev.* **2017**, *79*, 38–43. [CrossRef]
46. Huggins, J.L.; Raverty, S.A.; Norman, S.A.; Calambokidis, J.; Gaydos, J.K.; Duffield, D.A.; Lambourn, D.M.; Rice, J.M.; Hanson, B.; Wilkinson, K.; et al. Increased harbor porpoise mortality in the Pacific Northwest, USA: Understanding when higher levels may be normal. *Dis. Aquat. Organ.* **2015**, *115*, 93–102. [CrossRef]
47. Pearcy, W.G.; Fisher, P. Ocean distribution of the American shad (*Alosa sapidissma*) along the Pacific coast of North America. *Fish. Bull.* **2011**, *109*, 440–453.

48. Pugliares-Bonner, K.; Bogomolni, A.; Touhey, K.; Herzig, S.; Harry, C.; Moore, M. Marine Mammal Necropsy: An Introductory Guide for Stranding Responders and Field Biologists. *Woods Hole Oceanog. Inst. Tech. Rept.* **2007**. [CrossRef]
49. Williams, R.; Gero, S.; Bejder, L.; Calambokidis, J.; Kraus, S.D.; Lusseau, D.; Read, A.J.; Robbins, J. Underestimating the damage: Interpreting cetacean carcass recoveries in the context of the Deepwater Horizon/BP incident. *Conserv. Lett.* **2011**, *4*, 228–233. [CrossRef]
50. Moore, M.J.; Mitchell, G.H.; Rowles, T.K.; Early, G. Dead cetacean? Beach, bloat, float, sink. *Front. Mar. Sci.* **2020**, *7*, 333. [CrossRef]
51. Hasselman, D.J.; Hinrichsen, R.A.; Shields, B.A.; Ebbesmeyer, C.C. The rapid establishment, dispersal and increased abundance of invasive American shad in the Pacific Northwest. *Fisheries* **2012**, *37*, 103–114. [CrossRef]
52. U.S. Fish and Wildlife Service. Available online: https://www.fws.gov/fisheries/freshwater-fish-of-america/american_shad.html (accessed on 1 June 2020).
53. Lodi, L.; Hetzel, B. Rough-toothed dolphin, *Steno bredanensis*, feeding behavior in Ilha Grande Bay, Brazil. *Biociências* **1999**, *7*, 29–42.
54. Best, R.C.; da Silva, V.M.F. Inia geoffrensis de Blainville, 1817. *Mamm. Species* **1993**, *426*, 1–8.
55. Ronje, E.I.; Barry, K.P.; Sinclair, C.; Grace, M.A.; Barros, N.; Allen, J.; Balmer, B.; Panike, A.; Toms, C.; Mullin, K.D.; et al. A common bottlenose dolphin (*Tursiops truncatus*) prey handling technique for marine catfish (Ariidae) in the northern Gulf of Mexico. *PLoS ONE* **2017**, *12*, e0181179. [CrossRef]
56. Ralls, K.; Mesnick, S.L. Sexual dimorphism. In *Encyclopedia of Marine Mammals*; Perrin, W.F., Wursig, B., Thewissen, J.G.M., Eds.; Academic Press: San Diego, CA, USA, 2002; pp. 1071–1078.
57. Hohn, A.A.; Brownell, J.R.L., Jr. *Harbor Porpoise in Central California Waters: Life History and Incidental Catches*; IWC Report SC/42/SM47; International Whaling Commission: Cambridge, UK, 1990.
58. Norman, S.A.; Hanson, M.B.; Huggins, J.; Lambourn, D.; Calambokidis, J.; Cottrell, P.; Greene, A.; Raverty, S.; Berta, S.; Dubpernell, S.; et al. Conception, fetal growth, and calving seasonality of harbor porpoise (*Phocoena phocoena*) in the Salish Sea waters of Washington, USA and southern British Columbia, Canada. *Can. J. Zool.* **2018**, *96*, 566–575. [CrossRef]
59. Read, A.J.; Hohn, A.A. Life in the fast lane: The life history of harbor porpoises from the Gulf of Maine. *Mar. Mammal Sci.* **1995**, *11*, 423–440. [CrossRef]
60. Lockyer, C.; Kinze, C. Status, ecology and life history of harbour porpoise (*Phocoena phocoena*), in Danish waters. *NAMMCO Sci. Publ.* **2003**, *5*, 143–175. [CrossRef]
61. Gallagher, C.A.; Stern, S.J.; Hines, E. The metabolic cost of swimming and reproduction in harbor porpoises (*Phocoena phocoena*) as predicted by a bioenergetic model. *Mar. Mammal Sci.* **2018**, *34*, 875–900. [CrossRef]
62. Yasui, W.Y.; Gaskin, D.E. Energy budget of a small cetacean, the harbour porpoise, *Phocoena phocoena* (L.). *Ophelia* **1986**, *25*, 183–197. [CrossRef]
63. Gaskin, D.E.; Watson, A.P. The harbor porpoise, Phocoena phocoena, in Fish Harbour, New Brunswick, Canada: Occupancy, distribution, and movements. *Fish. Bull.* **1985**, *83*, 427–442.
64. Smith, G.J.D.; Gaskin, D.E. An environmental index for habitat utilization by female harbour porpoises with calves near Deer Island, Bay of Fundy. *Ophelia* **1983**, *22*, 1–13. [CrossRef]
65. Christiansen, F.; Víkingsson, G.A.; Rasmussen, M.H.; Lusseau, D. Female body condition affects foetal growth in a capital breeding mysticete. *Funct. Ecol.* **2014**, *28*, 579–588. [CrossRef]
66. Mukherjee, S.; Heithaus, M.R. Dangerous prey and daring predators: A review. *Biol. Rev.* **2013**, *88*, 550–563. [CrossRef] [PubMed]
67. McGovern, B.; Gridley, T.; James, B.S.; Elwen, S. Risky business? A note on repeated live strandings of common bottlenose dolphins (*Tursiops truncatus*) while foraging in a shallow water environment. *Mar. Mammal Sci.* **2019**, *36*, 305–314. [CrossRef]

oceans

Review

Harbour Seals: Population Structure, Status, and Threats in a Rapidly Changing Environment

Marie-Anne Blanchet [1,*,†], Cécile Vincent [2], Jamie N. Womble [3], Sheanna M. Steingass [4] and Geneviève Desportes [5]

1 Norwegian College of Fishery Science, UiT The Arctic University of Norway, 9219 Tromsø, Norway
2 CEBC, UMR 7372, CNRS, La Rochelle Université, 5 Allée de l'Ocean, 17000 La Rochelle, France; cvincent@univ-lr.fr
3 National Park Service, Glacier Bay Field Station, 3100 National Park Road, Juneau, AK 99801, USA; Jamie_Womble@nps.gov
4 Oregon Department of Fish and Wildlife, 7118 NE Vandenberg Ave, Corvallis, OR 97330, USA; Sheanna.M.Steingass@state.or.us
5 North Atlantic Marine Mammal Commission, POP 6453, Sykehusveien 21-23, 9294 Tromsø, Norway; genevieve@nammco.org
* Correspondence: marie-anne.blanchet@npolar.no
† Current Address: Norwegian Polar Institute, Fram Centre, 9296 Tromsø, Norway.

Abstract: The harbour seal (*Phoca vitulina*) is the world's most widely distributed pinniped species ranging from temperate to Arctic regions (30–78.5° N in the Atlantic, 28–61.2° N in the Pacific), but no detailed overview of the species status exists. The aims of this review are to (i) provide current information on the genetic structure, population status, and threats; (ii) review potential consequences of a changing climate; and (iii) identify knowledge gaps to guide future research and monitoring. Although the species is globally abundant, wide differences exist across the species' broad range. As climate warms, populations at the edges of the species' distributional range are likely to be more affected. The primary climate-related drivers include: (i) changes in weather patterns, which can affect thermoregulation; (ii) decrease in availability of haul-out substrates; (iii) large-scale changes in prey availability and inter-specific competition; (iv) shifts in the range of pathogens; (v) increase in temperature favouring the biotransformation of contaminants; and (vi) increased exposure to pollutant from increased freshwater run-off. Multiple anthropogenic stressors may collectively impact some populations. Coordinated monitoring efforts across and within regions is needed. This would allow for a spatially explicit management approach including population-specific responses to known stressors.

Keywords: harbour seal; *Phoca vitulina*; pinniped; distribution; population status; climate change; edge effect; knowledge gaps

Citation: Blanchet, M.-A.; Vincent, C.; Womble, J.N.; Steingass, S.M.; Desportes, G. Harbour Seals: Population Structure, Status, and Threats in a Rapidly Changing Environment. *Oceans* **2021**, *2*, 41–63. https://doi.org/10.3390/oceans2010003

Received: 3 November 2020
Accepted: 22 December 2020
Published: 5 January 2021

Publisher's Note: MDPI stays neutral with regard to jurisdictional clai-ms in published maps and institutio-nal affiliations.

1. Introduction

The harbour seal or common seal (*Phoca vitulina* Linnaeus 1758) is the most widely distributed pinniped in the northern hemisphere and ranges from temperate to Arctic regions. Due to its extensive range, nearshore coastal distribution, site fidelity, and high visibility, it is one of the most well-studied pinnipeds in the world [1,2]. Harbour seals use an array of habitats including bays, rivers, lakes, estuaries, intertidal habitats, sea ice, and icebergs in tidewater glacier fjords [3–8]. Harbour seals typically use solid substrates for birthing, nursing their young, resting and moulting, and can undertake extensive at-sea foraging trips lasting several days to weeks [9–11]. Harbour seals typically produce one pup each year in the late spring to early autumn depending upon the region [12], and mating occurs towards the end of the brief lactation period. Given that harbour seals occupy habitats that are in close proximity to human populations, the species has the potential to be exposed to a variety of anthropogenic activities including harvest, population reduction programs, coastal development, agricultural runoff and pollution, and interactions with fisheries [13,14].

Similar to other marine mammal species, harbour seals are expected to undergo changes in all or part of their range due to climate related changes which include ocean warming [15], ocean acidification, changes in the precipitation, decreases in sea ice and sea level rise. Collectively, these factors are altering the physical and biological environment occupied by marine species [16,17], and in many cases challenging their capacity to adapt [18–20]. It has been suggested that climate-related changes may be a predominant threat to pinnipeds, through changes in ecological processes especially in polar regions [20–26].

The harbour seal has the widest distribution of any coastal pinniped, ranging continuously from 28° to 78.5° of latitude north [2]. Given this broad geographic distribution the species encounters an extensive gradient of environmental conditions from temperate to Arctic regions. Thus, it presents a unique case study to understand the influence of changing environmental conditions on a single species. Here, we (i) review current information on the harbour seal genetic structure, population status, and threats; (ii) assess potential consequences of a changing climate; and (iii) identify knowledge gaps to guide future research and monitoring.

2. Distribution and Genetic Structure

In this review, we use "population" as a unit where virtually no gene flow would be expected within one generation and "stock" as a management unit. Hence one population can be composed of one or several stocks [27].

Harbour seal were previously recognized as five subspecies based on differences in morphological characteristics and geographic distribution. However, recent genetic analysis suggests three primary subspecies [28,29]. There are currently three recognized subspecies of harbour seal: the Atlantic harbour seal (*P. v. vitulina*, Linnaeus 1758), the Pacific harbour seal (*P. v. richardii*, Gray 1864), and the Ungava harbour seal (*P. v. mellonae*, Doutt 1942), which is endemic to a freshwater system in Canada [30] (Figure 1). Pacific harbour seals occur along the North Pacific Rim ranging from southern Japan, the western Aleutians and Bering Sea; Alaska; the Pacific coast of North America including British Columbia, Washington, Oregon and California (USA) to Baja California (Mexico). Atlantic harbour seals occur along the East Atlantic Ocean from Brittany (France) to the Barents Sea (Norway) including the British Isles, Iceland, and Greenland. In the western Atlantic, they range from New York (USA) to the Canadian Arctic. They are occasionally seen as far south as South Carolina (USA). Currently, the Atlantic and Pacific harbour seals are isolated from one another by the high Arctic coasts of Russia and Canada.

Pacific and Atlantic populations are sister clades with genetic differences within each subspecies at the local scale [28,31]. These differences occur between neighbouring colonies despite the absence of major geographical barriers [29,31–34] and are likely due to the species' relatively small scale movements at a maximum range of 300–500 km [35]. Genetic differentiation among groups of *P. vitulina* has been detected on a scale of only a few hundred kilometres in the northeast Pacific along their 16,000 km continuous distribution [29,31,36,37]. It appears that dispersal patterns of harbour seals are behaviourally restricted, and follow specific geographic features that likely limit gene flow between neighbouring regions [38]. This genetic variation is reflected in differences in phenology and morphology observed within the North Pacific Region following an axis along the Pacific coastlines and varying with geographic features including the continental shelf. Recent studies suggest that harbour seals in Japan originate from more than two lineages and secondary contacts between populations after long isolation [37,39]. In Alaska, harbour seals were previously managed as three stocks, but more recent evidence suggests structuring at a finer scale and twelve stocks are now recognized [40]. In the Atlantic Ocean, taxonomic schemes have recognized divisions between Greenland, North America and Europe [31]. Based on neutral genetic markers twelve genetically distinct populations across the north Atlantic Ocean are now identified [33]. The northernmost population on Svalbard in the Barents Sea is currently recognized as an independent, highly distinct genetic unit along with the small population in southern Greenland [41]. Surprisingly, despite low numbers and an apparent separation between the two populations in Greenland, the population in southern Greenland exhibits a high level of genetic diversity. It is as high as in the much larger population in Europe. This is also the case in the UK where stronger connections exist between the populations in the southeast of England and those in Europe, than between England and the neighbouring harbour seals from eastern Scotland [42]. Recently these two British populations have been further divided into four geographically distinct clusters [43].

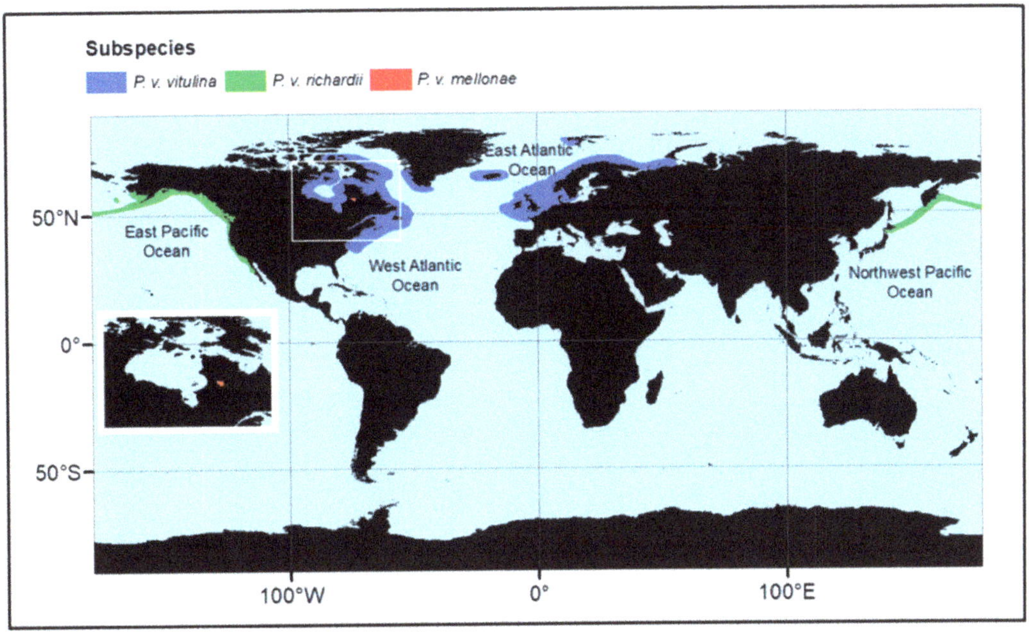

Figure 1. Distribution of harbour seals (*Phoca vitulina*). Colours correspond to each subspecies. The inset is a close-up on the region where *Phoca vitulina mellonae* is distributed. Modified from the International Union for the Conservation of Nature (IUCN) according to [2].

3. Status of the Main Populations and Current Threats

The worldwide population size of harbour seals is estimated between 610,000–640,000 individuals [44]. Although the global population trend is currently unknown, this species is listed as "Least concern" by the International Union for the Conservation of Nature (IUCN) red list [45]. However, given the broad geographic distribution, dramatic differences exist between subspecies, regions or populations in terms of minimum population estimates and population dynamics. Some populations are stable or increasing whereas others are experiencing declines leading to conservation concerns [21,44–47]. In addition, there are a wide variety of approaches that are used for monitoring, management, and conservation across the range of harbour seals.

3.1. Western Pacific Coast

Population dynamics of *P. v. richardii* in the western North Pacific region are not well documented due to an uneven distribution covering numerous remote islands [46,48]. In coastal regions of Russia, harbour seals occur on the Kuril Islands in the Okhotsk Sea and on the Commander Islands in the Bering Sea [49]. The Kuril Islands' population is thought to be around 3000 and seemed stable in the early 2000s [49]. The most recent count from the Commander Islands gives a total of 3344 individuals during the breeding season in July 2017 [50]. In the early 1990s a small population in the Kamchatka Peninsula was estimated at around 200 individuals [51], but no recent estimate exists. The harbour seal is listed in the Red Data Book of the Russian Federation and protected [50]. In Japan, harbour seals occur on the Pacific side of the island of Hokkaido and have decreased precipitously from the 1940s to only few hundred individuals in the 1970s [46]. Reasons for this decline are not clear but may include commercial hunting, bycatch in salmon nets, and interactions with coastal fisheries. In addition, the destruction of haul-out sites to improve substrates for commercial kelp production has likely contributed to this population's decline in the past [37]. However, recent trends show a yearly population growth rate of 4% and more than 1000 seals were recorded in the 2008 survey [37,46]. This is likely due to the cessation of the commercial seal harvest and the species' protection in Japan, although bycatch in salmon nets remains a concern in this region [37,46].

3.2. Eastern Pacific Coast

Harbour seals in the eastern north Pacific spans a diverse array of habitats across an extensive geographic range of over 8000 kilometres from the Aleutian Islands in western Alaska to Baja, California. Given this extensive range, the populations status varies between regions.

3.2.1. Alaska

Harbour seals range from Dixon Entrance in southeast Alaska to the Aleutian Islands and Bering Sea in southwestern Alaska. Although previously managed as three stocks, twelve stocks of harbour seals are currently recognized based primarily on genetics, movement data, and traditional ecological knowledge [40,52]. Harbour seals use a diverse range of habitats including beaches, sand bars, rocky islets, a freshwater lake, and icebergs that are calved from tidewater glaciers. Some of the largest aggregations of harbour seals in the world occur seasonally in tidewater glaciers fjords in Alaska, where they use ice that emanates from tidewater glaciers as habitat for pupping, moulting, and resting [53]. Although tidewater glaciers are naturally dynamic [54], the majority are retreating and thinning with unknown impacts on the seals that use iceberg habitat. A small, presumably isolated population of harbour seals (~400 seals) occurs in Iliamna Lake, a large freshwater lake that is connected to Bristol Bay by the Kvichak River [55]. Seals appear to be resident in the lake throughout the year and remain in the vicinity of cracks in the ice during winter [55,56]; however, the extent to which exchange occurs with seals in Bristol Bay is unknown [55].

Population status and their trajectories vary by region. Populations in the Bering Sea and Aleutian Islands were considered stable in the 1960s and 1970s [57–59]; however, surveys from the late 1990s documented precipitous declines [57,59,60]. Declines also occurred in the Gulf of Alaska [60–63]; however, more recent data suggests that some declines may have lessened [60,64]. In Glacier Bay, a tidewater glacier fjord and marine protected area in southeastern Alaska, precipitous declines in the number of harbour seals have occurred over the last 26 years [65–67]. Declines have also been documented in seabirds and other pinniped species and suggest that large scale changes in ocean climate and/or regime shifts [68,69] may have played a role [63,70–72]. Between 2014 and 2016, anomalously warm waters occurred in the eastern Gulf of Alaska due to a combination of a large warm water mass and a strong El Niño [73] which coincided with changes in lower and mid-trophic levels [74]. This also coincided with lower abundance estimates for harbour seals in Glacier Bay [68]. Collectively, given the extensive geographic range of harbour seals across Alaska, it is likely that a combination of larger-scale and local factors, such as variation in prey availability, predation, and/or habitat may have played a role in population trajectories over the last few decades [66,67,70,74,75].

3.2.2. British Columbia, Washington, Oregon, and California

Current population trends and abundance of harbour seals in British Columbia (B.C.) are assessed based on aerial surveys conducted from 1966 to 2014. Total abundance of harbour seals on the B.C. coast in 2008 was estimated to be 105,000 seals (95% CI: 90,900 to 118,900) [76]. Reconstruction of historical data indicates that the population was depleted during a period of commercial harvesting from 1879 to 1914, and subsequently maintained below natural levels by predator control programs until the early 1960s. The population was further depleted in the 1960s but now appears to have fully recovered [76]. Seal populations from coastal Oregon through Southern California seem to be generally increasing showing impressive recovery after discontinuation of historical state-financed bounties [77–79]. However, these populations seem to have reached now an asymptotic growth suggesting that they may be approaching carrying capacity [77,80]. Harbour seals in Washington state are divided into coastal and inland water stocks [77,81]. Their number approximates 30,000 individuals, although current data are not available. The primary population-level stressors likely include shifts in productivity and prey availability, organic pollutants, predation, and interaction with fisheries. In Oregon, harbour seals are estimated to be 10,000 individuals, based on aerial surveys in 2003 [77] and seem stable. Along the Oregon Coast, harbour seals occupy more than 90 haul-out locations including rocky shorelines, beaches, bays, estuaries, and outlying rocks [77,78]. They extensively utilize the continental shelf to forage, spending a great amount of time in highly productive areas with offshore seamounts and rocky substrates [38]. Current stock assessments treat the Oregon and Washington coastal stocks as one unit. However, seals in this region seem further subdivided based on geophysical barriers [38] such as Cape Blanco, where there is a narrowing of the continental shelf that separates the Central and Southern Oregon coasts. The California stock is estimated at 31,000 seals and appears stable in recent years [81]. The main limiting factors of population growth include changes in prey availability that are influenced by larger-scale inter-annual oceanographic processes such as El-Niño. In addition, a lack of undisturbed or available habitats, bycatch in commercial fisheries, and/or predation have been shown to negatively affect local harbour seal populations [82,83]. In Mexico, along the Pacific coast, harbour seals are found on several islands and the region represents the southern limit of the species' distribution [82,83]. The estimated number of seals is approximately 1000 individuals, but limited information regarding trends or threats exists [5].

3.3. *Western North Atlantic Coast*

Western Atlantic harbour seals range along the northern coast of Canada and the US from Baffin Island through North Carolina [84–86]. In Canada, three distinct units are recognized and include populations in Hudson Bay, Gulf of St Lawrence, and Sable Island [85,87]. A variety of survey methods have been used along the coast of Canada; and minimum population estimates range between 8000 and 12,000 seals, excluding Newfoundland [87]. In the US, coast-wide aerial surveys are primarily available for coastal Maine, with the last survey being conducted in 2012, providing a minimum population estimate of 75,834 seals (CV = 0.15) [88]. At Sable Island, the population has decreased dramatically since the early 1990s [89,90]. This decline appears to be due to a combination of shark-inflicted mortality and inter-specific competition with grey seals (*Halichoerus grypus*) [89,90]. The main threat to these populations is linked to interactions with fisheries through bycatch and reduced food availability [85,86]. A significant number of animals is caught each year in gillnets, bottom and mid-water trawls and small trap-nets, although it is unclear whether this translates to population-scale impacts [91,92].

P.v. mellonae is an endemic subspecies of harbour seals and the only that occurs throughout the year in freshwater. This population occurs along the Ungava peninsula in northern Quebec and has been isolated from harbour seals in the neighbouring Hudson Bay since the most recent glaciation, between 3000 and 8000 years ago. During winter, the seals are confined in ice-free areas and travel on snow between neighbouring lakes. They are genetically distinct from harbour seals in Hudson Bay and they also have darker pelage and a flatter skull. They typically breed earlier than individuals from the saltwater subspecies in the same region. The population size is small, estimated between 50–600 animals [30] which represents a threat to its viability. Additional threats to this population include the development of hydroelectric dams that could decrease areas of open water during winter and mercury contamination from fish [30]. Due to the limited adaptive capabilities of this small and isolated population, climate-related disturbances are also of concern [30]. In 2018, this population was designated as Endangered in Canada due to cumulative anthropogenic disturbances [92].

3.4. Greenland

Historically, harbour seals were widely distributed throughout Greenland although not numerous. In recent years they became rare in most areas and in some cases extinct [3]. Currently, their status is "critically endangered" on the Greenlandic Red List and they are protected from hunting throughout the year. Two primary populations exist; one located on the west coast of Greenland and a second limited to the southeastern tip of the island. The population along the west coast is closely related to the West Atlantic populations whereas seals from the southern population are genetically closer to the Icelandic and Svalbard populations [41]. Current population estimates are not available; however, there is some historical information based on catch and skin statistics [47]. The number of seals has declined rapidly in West Greenland since the 1950s, most likely due to unstainable hunting pressure. On the other hand catches in the southern part of Greenland remained stable between 1960–1980 and there were active breeding sites through the 1990s [3,93]. Harbour seals in Greenland were believed to number fewer than 1000 in the mid-2000s, representing one-third, of the estimated population size in the 1950s [47]. The primary threat to these populations is their small size and apparent separation, which makes them particularly vulnerable to stochastic events.

3.5. Svalbard

The Svalbard harbour seal constitutes the northernmost population of this species. It is genetically distinct from neighbouring populations and apparently isolated [4,94–96]. It is the only population to inhabit a true Arctic environment throughout the year [95,96]. Adults and juveniles are observed along the west coast of Spitsbergen throughout the year, with the northernmost record being as far north as 80.5° N [10,97]. They are very rarely observed past the southern tip of the island. Along the eastern coast of Svalbard, their distribution is limited by the occurrence and the thickness of sea ice [10,98–100]. Surveys were conducted in 2009 and 2010 and estimated a total of 2000 seals [4]. Individuals from this population tend to be shorter and more rotund compared to their southern counterparts suggesting adaptations to a colder environment [99]. The Svalbard population exhibits a high degree of sexual dimorphism compared to more southerly populations, with adult males being significantly heavier and longer than adult females [4]. The longevity of Svalbard harbour seals seems somewhat shorter than in other populations. The apparent lack of individuals older than 16 years is surprising given limited the human-seal interactions and the absence of acute source of mortality from epizootic outbreak. This skewed demographic distribution, with few older individuals might be linked to pressure from terrestrial predators such as polar bears (*Ursus maritimus*), or marine predators, such as killer whales (*Orcinus orca*) or Greenland sharks (*Somnius microcephalus*) [100]. Recent studies on contaminants show that this population is exposed to a wide variety of pollutants; however, measured levels suggest that these are not an immediate threat to their health [101]. The main threat to this population is linked to its small size and low genetic diversity which could reduce its resilience to stochastic events such as oil spill or disease outbreaks [41]. Presently, this population is red-listed in Norway and protected from exploitation.

3.6. Iceland and Faroe Islands

The harbour seal is the most abundant pinniped in Iceland [102–104] and has likely been exploited by humans since the settlement of the region, but nowadays, there is no commercial harvesting. In an effort to subsidise the seal industry and to control the incidence of roundworm (*Pseudoterranova* sp.) in commercial fish, a bounty program was introduced between 1982 and 1989 resulting in high levels of adult mortality [104]. Until 2018, harbour seal culling remained subsidised by the angling industry to protect salmonids from predation as this fishery is economically in Iceland [105–107]. Regular aerial surveys since the early 1980s show that the population has decreased dramatically from approximately 33,000 individuals in 1980 to 12,000 in 2006 [107] and to about 7700 animals in 2016 [108]. The latest surveys in 2018 indicates that the population has now increased to about 9400 individuals [108]. However, this number remains 21% lower than the 2006 government issued management objective for a minimum population size of 12,000 animals [108]. Reasons for this decline are not well understood but may include over-harvesting, changes in prey availability, bycatch, environmental changes, and anthropogenic disturbances [108]. In the Faroe Islands, harbour seals were common in sheltered fjords, and likely more common than grey seals until the mid-19th century, when the species was extirpated through extensive harvesting [109]. Since then, harbour seals have been observed during bounty hunts, from 1889 to 1891 and again from 1963 to 1967 when one and four seals were caught respectively, in the southernmost part of the Southern Island [109]. Since then, only three observations of seals have been made in this region in 2001, 2005 and 2019 [110].

3.7. Continental Europe

Harbour seals in continental Europe are distributed over a wide latitudinal range from France (48° N) to northern Norway and the Murman peninsula (70° N) [111,112]. Through most of its historical distribution, the species has been harvested for fur or meat [32,113]. In addition, seals were also considered as competitors by fisheries, and long-term bounty programs depleted several populations until the mid-1970s [14,51,114]. Following the reduction in hunting pressure most populations in continental Europe started to recover, although in an unequal way. This recovery has, however, been hampered by two consecutive epidemics of Phocine distemper virus (PDV). These epidemics that swept through most of the European populations causing the death of 230,000 and 30,000 seals in 1988 and 2002, respectively, which represented over half of the total population [9,115].

3.7.1. Northern Europe

Harbour seals occur along the northernmost coast of Europe from northern Norway to the Eastern Murman coast in Russia where the easternmost breeding colony of *P. v. vitulina* subspecies is found [116,117]. The latest population estimate in the late 1990s ranged from 400–500 seals [116]. The main threat to this remote population is human disturbances at breeding sites, poaching, bycatch and shooting at salmon nets. Currently, the species is listed in the Red List of the Murman area and hence banned from harvesting. In mainland Norway, a system of quotas that was established after the culling period 1980–1987 regulates the hunt [115]. Quotas were increased substantially from 2003. In 2010, a management plan for harbour seals was implemented, with quotas aimed at maintaining the population at target level and surveys occurring every five years. The estimated number of seals in mainland Norway is approximately 7500 individuals during the 2011–2015 period. The number of seals appears stable compared to previous periods (2003–2006 and 1996–1999) [118]. Harbour seals along the Norwegian coast are threatened mainly by fisheries through bycatch and interactions with fish farms [114], although shooting at fish farms was prohibited in November 2019 (executive order FOR-2019-11-28-1593).

In southern Scandinavia and southern Baltic, harbour seals are divided into four populations, the Limfjord, the Kattegat, the Southern Baltic Sea and the Kalmarsund [119]. Heavy hunting pressure brought these populations of harbour seals to a historical low in the 1920s [120]. In the late 2000s the estimated number of seals in Southern Scandinavia and

Baltic was estimated between 26,350–33,450 individuals but has subsequently increased. Significant number of seals died in mass mortality events in 1998 and 2002, due to PDV, in 2007 due to an unknown cause, and in 2014 due to avian influenza [115,121–123]. Currently all populations in Southern Scandinavia are increasing and, except for the Kalmarsund one, seem to have reached or be approaching carrying capacity [121,123]. The Kalmarsund population a small, genetically isolated population has been increasing at 8% per year over the period from 2003 to 2016 reaching 1000 seals in 2014 [124]. The main pressures affecting seals in southern Scandinavia include bycatch, shooting, exposure to contaminants, human-related disturbances at breeding sites, and interspecific competition with grey seals [121,123,124]. However, given the recent increasing population trends it is difficult to assess whether these sources of mortality have population-level effects. In the Baltic region, infertility due to organohalogen pollution was identified as a problem for harbour, ringed and grey seals in the 1970s but since that time these effects seem to have decreased [123,125].

3.7.2. Southern Europe

In southern Europe, from the Netherlands through France, harbour seal populations appear to be increasing despite their proximity to human activities and heavy exploitation of the coastal areas that they inhabit. The population in the Wadden Sea is estimated to be between 25,000 and 31,800 individuals and has shown a quick recovery after two PDV epizootics [12,14,126]. The most recent total population estimate in 2019 was 40,800 in the Danish, Dutch and German Wadden Sea [126]. Increasing exploitation of coastal areas and shipping in the North Sea represent the primary threats to the Wadden/North Sea population. In particular, offshore wind farms may have the potential to interfere with foraging and migratory behaviour; although no impact studies have been conducted [127]. In France, where only three colonies exist, harbour seals have been completely protected since 1995. The three colonies show increasing trends with a minimum population estimate of 830 animals during the moult [128]. Observation of individuals from neighbouring colonies indicate some exchange with populations from the southern part of the United Kingdom (UK) and the Netherlands.

3.8. United Kingdom and Ireland

Approximately 40% of the European harbour seal population occurs in the UK with the majority around the coast of Scotland [129–135]. The most recent estimate in 2016 was 43,450 (95% CI: 35,550–57,900) seals [129]. Colonies on the northwest and southeast coasts appear to be stable or increasing [130,131]. Scottish colonies have experienced dramatic declines, especially on the east coast (Orkneys and Shetland) where populations have decreased by 85% between 2000 and 2010 [133,134]. Reasons for these widespread declines are yet unclear, but research efforts are currently focussed on competition with grey seals, predation from killer whales, and exposure to toxins from harmful algal blooms [133,134,136]. Recent studies reveal that harbour seals in declining colonies are significantly more exposed to harmful algal toxins, such as domoic acid and saxitoxins, which may be a contributing factor to the observed declines [136]. It is interesting to note that although the population in the Wash, England, is increasing, the rate of increase is still lower than in the neighbouring population in the Wadden Sea [130]. Harbour seals are relatively common in coastal waters of the Republic of Ireland at the edge of the species' range in Northwest Europe. Although haul-out sites cover the entire coast, scarce information is available regarding the population's trajectory [137–140]. A survey conducted in the early 2000s yielded a minimum population estimate of approximately 2,905 individuals [139], but no population trend is currently available due to the lack of historical and current data. Anecdotal data indicates increasing numbers of animals in southwest Ireland. Seal predation and damage to fishing gear is currently not monitored, but fishers and aquaculture operators are still licensed to shoot seals interacting with fishing equipment. An additional threat to this population includes bycatch, especially in the vicinity of major colonies [141,142].

4. Environmental Changes and Potential Consequences on Harbour Seals Populations

4.1. Increased Temperatures and Extreme Weather Events Affect Haul-Out Patterns

The average global land and ocean surface temperature for January–August 2019 was 0.94 °C (1.69 °F) above the 20th century average [143]. Ocean warming dominates the increase in energy stored in the climate system [18]. Changes in weather patterns have occurred over the last 50 years with an increased probability of extreme weather events such as heat waves, storms and large amount of precipitations [144]. These changes in the physical environment are expected to directly and indirectly influence marine mammals, especially amphibious species such as pinnipeds that use land, aquatic and ice environments. Harbour seal haul-out patterns are directly influenced by water and air temperature because thermoregulation is energetically costly [145–147]. Thus, in cold environments, harbour seals must continually thermoregulate to mitigate heat loss at sea, on land and ice [101,145–147]. For example, in Svalbard, at the northern limit of their distributional range, harbour seals typically spend more time at sea, during stormy weather conditions, even if the temperature is high because of the wind chill effect [101]. Given that the frequency of storms is predicted to increase in the Svalbard archipelago, it is expected that haul-out patterns could change and influence harbour seal energy budget. At the southern limit of their distributional range, harbour seals face the opposite problem. Hyperthermia is observed in juvenile seals at an ambient air temperature of 35 °C [147], a temperature that is easily reached in Southern California, Mexico, and France. This is especially critical for new-born pups that haul out for the majority of their time in the first weeks after birth [147]. Adult animals are also subject to overheating which can be a limiting factor in hauling out. Thus, a northward shift in distribution can be expected if southern populations cannot cope with high temperatures at least during the breeding and moulting periods. For northern populations, an increase of water and air temperatures may actually decrease thermic stress during the winter months and be an advantage [101].

4.2. Changes in Physical Habitat Affect Distribution Patterns

Sea ice cover in Arctic and sub-Arctic regions has dramatically decreased in recent decades, reaching historic minimums in 2007 and 2012 [18,148,149]. Throughout most of their range, harbour seals are not typically associated with sea ice. They even tend to avoid areas with thick ice where they cannot maintain breathing holes and are at risk of predation by polar bears [10,150]. As such, decreasing sea ice may increase the available habitat for harbour seals in seasonally ice-covered areas such as Svalbard, the western Hudson Bay, the St Lawrence Estuary or Greenland [150–152], which in turn could result in a northwards range expansion. However, other features, such as bathymetry, may still limit their expansion, as harbour seals are generally considered relatively shallow divers [100,153,154]. Harbour seals occasionally use ice floes and land-fast ice as a resting platform close to foraging sites during the winter like on Svalbard. Whether the reduced availability of such platforms would be detrimental for harbour seals is unknown [67,99,117]. In the Murman region, it has been suggested that hauling out on ice may reduce the risk of predation from land-based predators [117]. In south Greenland, low inflows of drift ice have resulted in abnormally high catches of harbour seals because drift ice used to shelter the seals' terrestrial haul-outs from the hunters [3,47].

In southeastern and southcentral Alaskan fjords, glacier ice and icebergs that emanate from tidewater glaciers are important habitat for harbour seals. However, the majority of tidewater glaciers are thinning and retreating [155,156]. It is not known whether the decrease of ice as a platform will affect harbour seals during critical life history events such as pupping and moulting [157,158]. The use of glacier ice habitat may confer several benefits including reducing the risk of predation, disease and pathogen transmission and providing a stable platform for nursing young that is not subject to tidal inundation [158,159]. Thus, reduction in the availability of glacier ice as habitat could potentially have population-level consequences [160]. Additionally, in Arctic and subarctic regions harbour seal distribution and niche often overlap with the ones of ringed seals and harp seals (*Pagophilus Groenlandi-*

cus). This could lead to interspecific resource competition and changes in predator-prey interactions, if harbour seal distributional range expands northwards and the species becomes more abundant [158,159,161,162].

4.3. Large-Scale Oceanic Events and Changes in Community Structure Affect Foraging

Patterns of the North Atlantic Oscillation (NAO) and the El Niño-Southern Oscillation (ENSO) have contributed to major variations in climate worldwide and directly or indirectly influence animal populations [163,164]. Strong ENSO events and cyclic patterns of the NAO have been linked to failed reproduction and decreased attendance at haul out sites in several pinniped species through changes in prey abundance and distribution [80,87,163]. Population trends in harbour seals in Alaska generally follow those of sympatric Steller sea lions (*Eumetopias jubatus*) and northern fur seals (*Callorhinus ursinus*), suggesting drivers linked to large-scale regime shifts [69,164]. On the west coast of North America, in particular Oregon and Washington, a recurrent pattern of seasonal areas of upwelling-driven hypoxia and anoxia on the continental shelf has caused die-offs of invertebrates and redistribution of many important preys including schooling fish and bottom fish. These deoxygenation events are thought to have secondary effects on the foraging efficiency of air breathing predators, such as harbour seals, as their prey experience physiological stress and habitat compression [165,166].

Changes in community structure can affect predators foraging patterns and diet composition. For example, large changes have been observed in the Barents Sea communities [167–171] due to dramatic increase in the influx and temperature of Atlantic water. This boreal water mass enters the Arctic Ocean and changes the characteristics of the Barents Sea/Fram Strait region from Arctic to Atlantic [167–171]. The range of Atlantic cod (*Gadus morhua*) has expanded northwards and they may even occupy areas of the Arctic continental shelve threatening local species such as polar cod (*Boreogadus saida*) [171]. Harbour seals on Svalbard have now shifted to feeding mainly on Atlantic species such as the Atlantic cod and haddock (*Melanogrammus aeglefinus*) [172,173]. The seals react to upwelling phenomenon that brings Atlantic water onto the West Spitsbergen shelf and they likely target associated Atlantic fish species under these events [98]. Predicted increase influx of Atlantic water in this region in combination with decreased sea ice are likely going to favour the growth and geographic expansion of the harbour seal population. However, diet changes might come at a fitness cost in other regions such as in Scotland [174]. Harbour seals were in poorer body condition during years of low herring (*Clupea harengus*) abundance when switching to alternative preys such as gadoids and showed signs of fish-induced anaemia [175]. In the past 40 years in the North Sea major climate-induced regime shifts have caused changes in community structures with effects rippling through the entire trophic chain up to top predators [176,177]. Such large-scale regime shifts are predicted to increase in the future and will disrupt local conditions and ecological relationships. Generalist top predators such as harbour seals have a flexible and broad diet which allow them to switch between several trophic niches if they can cope with the physical environment. Such species have the potential to establish in new areas, such as the high Arctic [178], and often have a competitive advantage over true Arctic species, such as ringed seals. Changes in community structure are not limited to prey but may also involve predators. For example, killer whale sightings have been more frequent in the Canadian Arctic, suggesting that the species is becoming more established in the region at least seasonally [179,180]. Harbour seals are regularly consumed by killer whales and the spatial overlap between these two species in the Arctic is likely to increase in the future. Polar bears are also known to prey upon harbour seals in the Hudson Bay and Svalbard. In these regions, the decrease in ringed seals abundance, the polar bear main prey, might increase the predation on harbour seals (Kovacs and Lydersen pers. comm.).

4.4. Shift in Pathogen Ranges May Affect Populations' Trajectories

Warmer air and water also affect the susceptibility of harbour seals to infections by pathogens [181,182]. The susceptibility of a population to pathogens depends on several factors. They include pathogens' geographic range, altered host-parasite dynamics, changes in life cycle, increased virulence, and unpredictable patterns of diseases [183–185]. A warming environment means that minimum temperature thresholds that acted as limiting factors for parasite and bacterial survival and growth are raised, leading to increased environmental prevalence of free-living pathogens [185]. In particular, overwinter survival of pathogens or parasites can dramatically increase their range expansion, density-dependent transmission, reaching new host populations or even species [186–191]. Pathogens are more likely to be transmitted to immunologically naïve host populations via a range of mechanisms, including: invasive species; species and population overlap; shifts in prey consumption; seasonal migrations; and shifting habitat availability [186,189,191]. In addition, climate-mediated physiological stresses and exposure to environmental pollutants have been shown to compromise host immune function and thus increase the clinical occurrence of opportunistic diseases [192]. Epizootic viral diseases causing mass mortality in harbour seals have been mainly reported from Europe and the USA [115,193,194]. In 1988, 60% of the North Sea harbour seals died during an outbreak of PDV followed by a subsequent outbreak in 2002 [115,193]. Above-average mean monthly air temperatures in Europe and an increase in density of hauled-out seals have been linked to mass mortality events [194]. During both outbreaks, the first cases were reported in the late spring when harbour seals start hauling out in larger groups for the breeding period, which is likely to enhance rates of transmission of density-dependent diseases [194]. In Arctic regions where harbour seals haul-out on ice, the disappearance of this platform might force the seals to haul-out on land in denser aggregations as observed in Pacific walrus. This will potentially increase the risk of transmission of infectious diseases and parasites. In the particular case of the PDV, grey seals have been suggested to be vectors infecting geographically distinct harbour seal populations [115,194]. Although harbour seal populations north of 65° N have not been affected by these recurrent epidemics, grey seals could also expand their distributional range northwards and therefore be in contact with immunologically naïve northern populations of harbour seals such as in Greenland [195]. Such epidemic outbreaks could potentially have a disastrous effect on small, genetically distinct populations with limited immunological robustness that might not be able to withstand a great loss of individuals, such as the Svalbard and Greenlandic populations [196]. Antibodies for PDV have been detected in harp seals from Canada, Greenland and the Barents Sea meaning this species might act as a reservoir for the virus. PDV antibodies have also been detected on the Pacific side in Kuril harbour seals where the prevalence is high (up to 100% in some sites) and the seals haul-out in dense groups [197]. Five species of seals (harbour, spotted (*Phoca largha*), ringed, ribbon (*Histriophoca fasciata*) and bearded (*Erignathus barbatus*)) overlap in this region, increasing risks of inter-specific transmission [197]. In addition, PDV has been detected in sea otters (*Enhydra lutris*) [198], which likely creates a potential threat to harbour seals and other pinnipeds that overlap with sea otters in the North Pacific [198]. Along the north-eastern coast of the US, harbour seal mass mortality events have been attributed to the Influenza A virus which occurred several times in the early 1980s. Thus far, this virus has only been isolated from harbour seals from the east coast of the US but is thought to be transmitted via pelagic birds which can cover large areas and could potentially infect other populations of marine mammals [190]. Additionally, pinnipeds that inhabit nearshore regions near human settlements and have a semi-aquatic lifestyle will likely be at increased risk of pathogen exposure [186,187,199,200]. A variety of pinniped-related parasites have begun to expand their range mainly northwards under the influence of environmental parameters [200,201]. In one example of this, the obligate intercellular parasite *Toxoplasma gondii* has appeared in the Arctic food chain and the transmission path and complete lifecycle of this organism in the Arctic environment is still not clear [200]. Warmer seas have resulted in influxes of temperate marine species that could serve as vector for this parasite in the Svalbard Archipelago, even in the absence of the terrestrial definitive host. This might explain the high prevalence of the parasite in top predators in this region [200]. In the Kuril harbour seal, the increase of river runoff due to heavy precipitation has been linked to the appearance of oocysts of the protozoal endoparasites *T. gondii* and *Neospora caninum* in the marine ecosystem [199]. Warmer local sea temperatures have, in addition, enhanced survival of the parasite [197]. Although these parasites are not directly associated with mortality in harbour seals, an increased parasite burden can compromise the host's immunological function resulting in increased susceptibility to other pathogens or environmental stressors [201]. Mass mortality events in marine megafauna can lead to dramatic changes in abundance of lower trophic level species and community structure, playing an important role in shaping population dynamics and trajectories. Populations recovering from mass mortality events have a very different structure compared to the initial populations which renders difficult their management [14]. These abrupt changes in population structure may mask additive or synergistic drivers, such as anthropologic disturbances or climate-related changes.

4.5. Increased Anthropogenic Disturbances May Affect Pristine Populations

As Arctic sea ice declines, new trans-Arctic shipping routes are being established, increasing vessel traffic. This will result in noise and chemical pollution [202,203]. For harbour seals, disturbances have the greatest effect near haul-out sites and during the reproductive season when they tend to spend more time ashore. In particular, the northern populations in Svalbard or Greenland could be disproportionately impacted while Southern populations might be more resilient to anthropogenic factors to which they are already exposed.

Pollution loads are very different among harbour seal populations. Southern populations in close proximity to human settlements, typically carry heavier pollutant loads compared to their northern counterparts. Heavy contaminant loads in marine mammals are associated with compromised immune systems, hormonal disruptions and increased parasite burden [204–206]. Although few major sources of pollution typically originate in the Arctic, this region is nevertheless exposed to pollutants through atmospheric and marine transport and freshwater runoff. Climate-related changes may affect these pathways and could therefore affect the exposure in Arctic regions. There is also compelling evidence that increasing temperature could be deleterious to pollutant-exposed wildlife through alterations in the biotransformation of contaminants [206]. Biological pollution is also an emerging issue with an increased presence of terrestrial pathogens in the marine system. This is due partly to an increase of anthropological use of coastal areas, but also to the increase of watershed runoff after bouts of increasingly dramatic precipitation events. For example, significant amounts of faecal coliform bacteria have been detected in harbour seals living near human settlements [206]. Biotoxins released from marine phytoplankton blooms have been recently found in harbour seals from Scotland [136] and linked to a possible decline in local populations. These harmful algal blooms have globally increased in distribution and intensity since the 1980s and new areas can be exposed such as the Norwegian coast [207].

5. Management and Knowledge Gaps under Rapidly Increasing Environmental Changes

5.1. Current Management Framework

Harbour seals are under a wide array of conservation status and management regimes across their broad geographic range. The species is listed as "Least Concern" on the global IUCN Red List, as it is very widely distributed and the total population size numbers in the 600,000 [45,208,209]. The Eastern Pacific subspecies is either stable or increasing in most of its range [210], while trends remain unknown for the Atlantic subspecies. Both of these subspecies are listed as "Least Concern" in regional Red Lists [208,209]. Some smaller, distinct populations are locally listed as "Endangered" (Canada endemic Ungava seal), "Critically Endangered" (Greenlandic and Icelandic populations), or "Vulnerable" (Japan, Svalbard, Russian Federation, and Kalmarsund populations). Bounty systems were historically in place in Canada, the United States, Norway, and Iceland. They aimed at controlling local harbour seal populations in areas where fisheries and angling took place to reduce competition with fisheries and, in one case, to reduce the incidence of roundworm in commercial fish. Presently, there are no bounty systems remaining in place. However, harbour seals may be shot anytime for protection of fishing operations [130] in Scotland (under license), England and in Wales. The shooting of seals at fish farms was forbidden in 2019 in Norway and in 2020 in Canada. The harvest of harbour seals is allowed, but seasonally and/or is quota regulated, in England, Wales and Norway. In Iceland, harbour seals may be hunted under a special licence for traditional use.

Currently, harbour seals are protected from hunting:

- Under the Marine Mammal Protection Act in the USA (enacted in 1972) throughout its range, although subsistence hunting and traditional use are permitted for coastal Alaskan natives.
- By the Marine Mammal Regulations under the Fisheries Act in the non-Arctic part of Canada (since 1967 for the Pacific population and 1970 for the Atlantic population), while in the Arctic, subsistence hunting is permitted for both the marine and fresh water subspecies. The species is not managed in Nunavut.
- In Greenland (since 2010).
- In Iceland (since 2019).
- In Svalbard (since 1970s).
- Under the EU Habitats and Species Directive 1992 (Council Directive 92/43/EEC on the Conservation of natural habitats and of wild fauna and flora). The species is listed in Annex II (species requiring the designation of special areas of conservation, SAC, or marine protected areas, MPAs) and V (species whose taking from the wild can be restricted by European law). The monitoring of their population abundance and distribution is requested under the Marine Strategy Framework Directive (MSFD).
- In the Russian Federation (since 1975).
- In Japan under the Wildlife Protection, Management and the Hunting Law (since 2003) with some specific local population control plans.

The harbour seal is not listed in the appendices of the Convention on International Trade in Endangered Species of Wild Fauna and Flora (CITES). It is cited in Appendix III (Protected fauna species) of the Berne Convention. The Baltic and Wadden Sea populations are listed on Appendix II of the Convention of Migratory Species (species that need or would significantly benefit from international co-operation). The Baltic population is coordinated by the HELCOM seal Expert Group.

5.2. Knowledge Gaps

As environmental conditions change, up-to-date baseline data is important to inform conservation and management of harbour seals.

Genetic studies at local scale are still lacking in many regions. Further genetic sampling will facilitate stocks assessment and understanding of population dynamics at a regional scale. Some stocks, including several genetically distinct populations are still managed as single units even if some very contrasting population trends are apparent. More information on the degree of immigration and interbreeding between populations is also needed in some areas, especially if one of the populations is small, for example, between the Limfjord population and the Wadden Sea population. In this context, movement studies (biotelemetry) are useful for understanding the spatial distribution of individuals across the annual cycle and for understanding overlap with potential threats (see for example [11,38,98]).

Health and Disease Monitoring aiming at establishing baselines for health parameters and disease status, identifying causes of death as well as isolation and characterization of infectious agents are still lacking. However, some local initiatives exist [123,187]. Collecting data on health parameters, distribution, epidemiology and effects of pathogens is essential to understand the impacts of pathogens range shift [184]. Since the relatively recent discovery of the PDV in 1988, this virus has become the most ecologically significant pathogen in harbour seals [193]. Some populations have not been tested for the prevalence of the virus, rendering difficult the monitoring of a range expansions of this pathogen in relation to movements between colonies of harbour seals and of other pinnipeds and demographic impacts. The epidemics of the virus are still unclear in the northeast Atlantic especially with respect to potential reservoir species such as harp and grey seals, which are sympatric with harbour seals.

Bycatch is recognised as a source of mortality in many areas, e.g., [108,141,210]; however, estimates and observer programs are limited in most regions.

Cumulative effect of stressors represents a key knowledge gap, especially as effects might not be simply additives but synergistic. Harbour seal populations are impacted at different rates by a variety of stressors and exhibit contrasting trends even at small regional scales. Comparing populations' parameters relative to the dynamics of their stressors would allow to better understand the range of potential responses. It would help informing the degree of harbour seals biological and behavioural plasticity and in turn populations' resilience and adaptive potential.

Survey efforts varies across the distributional range. In some areas, such as in Alaska, surveys occur regularly, whereas in other regions monitoring efforts have not been conducted recently. The remote nature of many of the population creates logistical challenges for population monitoring. However, with expected environmental changes and increased anthropogenic stressors, it is essential to have up-to-date population estimates and reliable trends in abundance. This is especially important for the smallest populations with low genetic diversity and populations at the edges of the species range. When possible, coordinated and consistent monitoring methods between regions would be useful. They would facilitate comparative studies and thus increase the value of data from local populations, particularly for populations of conservation concern.

5.3. *Adaptive Management of Harbour Seals Following a Precautionary Approach*

Climate and environmental changes impose growing pressure on global biodiversity, which requires that managers have access to up-to-date information on ecosystems to make timely and informed decisions. Hence, consistent monitoring is an essential aspect of informed management and conservation. In some regions, monitoring programs remain largely uncoordinated, limiting the ability to monitor, understand and respond effectively to trends. Marine mammals are prime sentinels of marine ecosystem changes because they integrate and reflect ecological variation across large spatial and temporal scales. Spatially explicit management recommendations are needed to support the resilience of (sub)populations at all scales. For example, in Japan, the population is generally increasing, but some previously depleted sites have not been recolonized [46]. This local specificity increases the probability of local extinctions through stochastic events. Small populations, particularly those at the edge of the distributional range, i.e., likely close to their adaptation capabilities, are particularly vulnerable. Hence, future objectives must give a particular attention to small entities and ensure that anthropogenic activities do not jeopardize their future persistence. It is also crucial that target population level objectives are based on biological criteria [118]. Adaptive management based on regular monitoring is needed more than ever under the current rate of environmental changes.

6. Conclusions

This review summarizes wide disparities in populations and conservation status of harbour seal populations across their broad geographic range. These disparities are not surprising given the extensive geographic range that harbour seals occupy. Large differences also exist with respect to the level of information available. Some populations in the UK, southern Europe, southern Scandinavia, Western Canada and Alaska are well-monitored, while recent data are lacking for Eastern Canadian, Greenlandic, Icelandic, Russian and Japanese harbour seals. The European populations are still recovering after massive epizootic events that wiped out a large number of individuals rendering its difficult the estimation of true population trends and predict their fate once at carrying capacity. Climate-related change will likely play a significant role in observed and future trends in population dynamics and will also likely be synergistic with direct human disturbances such as bycatch, pollution and exploitation of coastal areas. Cumulative stressors might be further detrimental to already vulnerable populations, and populations at the edge of the range may have limited adaption capabilities. Climate predictions indicate a unilateral warming of the atmosphere and oceans. This is already causing a northward shift of the distributional range of numerous species likely including the harbour seal. Harbour seals may expand northwards, and the southernmost populations may possibly be lost permanently. Ultimately, to facilitate and further complete our understanding of harbour seal populations trends, it will be necessary to have a more integrated approach to ecological monitoring that also includes monitoring associated with bottom-up (e.g., oceanography and mid-trophic levels) and top-down (e.g., predation) processes that are known to influence harbour seal populations across their broad geographic range.

Funding: This research received no external funding.

Conflicts of Interest: The authors declare no conflict of interest.

References

1. Shaughnessy, P.D.; Fay, F.H. A review of the taxonomy and nomenclature of North Pacific Harbour seals. *J. Zool.* **1977**, *182*, 385–419. [CrossRef]
2. Teilmann, J.; Galatius, A. Harbor Seal. In *Encyclopedia of Marine Mammals*, 3rd ed.; Würsig, B., Thewissen, J.G.M., Kovacs, K.M., Eds.; Academic Press: Cambridge, MA, USA, 2018; pp. 451–455.
3. Rosing-Asvid, A.; Teilmann, J.; Olsen, M.T.; Dietz, R. Deep diving harbor seals (*Phoca vitulina*) in South Greenland: Movements, diving, haul-out and breeding activities described by telemetry. *Polar Biol.* **2020**, *43*, 359–368. [CrossRef]
4. Merkel, B.; Lydersen, C.; Yoccoz, N.G.; Kovacs, K.M. The World's Northernmost Harbour Seal Population—How Many Are There? *PLoS ONE* **2013**, *8*, e67576. [CrossRef] [PubMed]
5. Chávez-rosales, S.; Gardner, S.C. Recent harbour seal (*Phoca vitulina richardsi*) pup sightings in Magdalena Bay, Baja California Sur, Mexico. *Aquat. Mammals* **1999**, *25*, 169–171.
6. Boveng, P.L.; Bengtson, J.L.; Withrow, D.E.; Cesarone, J.C.; Simpkins, M.A.; Frost, K.J.; Burns, J.J. The abundance of harbor seals in the Gulf of Alaska. *Mar. Mammal Sci.* **2003**, *19*, 111–127. [CrossRef]
7. Vincent, C.; McConnell, B.J.; Delayat, S.; Elder, J.-F.; Gautier, G.; Ridoux, V. Winter habitat use of harbour seals (*Phoca vitulina*) fitted with Fastloc^{TM}GPS/GSM tags in two tidal bays in France. *NAMMCO Sci. Publ.* **2010**, *8*, 285. [CrossRef]
8. London, J.M.; Hoef, J.M.; Jeffries, S.J.; Lance, M.M.; Boveng, P.L. Haul-out behavior of harbor seals (*Phoca vitulina*) in Hood Canal, Washington. *PLoS ONE* **2012**, *7*, e38180. [CrossRef]
9. Dietz, R.; Teilmann, J.; Andersen, S.M.; Rige, F.; Olsen, M.T. Movements and site fidelity of harbour seals (*Phoca vitulina*) in Kattegat, Denmark, with implications for the epidemiology of the phocine distemper virus. *ICES J. Mar. Sci.* **2013**, *70*, 186–195. [CrossRef]
10. Blanchet, M.-A.; Lydersen, C.; Ims, R.A.; Lowther, A.D.; Kovacs, K.M. Harbour seal *Phoca vitulina* movement patterns in the high-arctic archipelago of Svalbard, Norway. *Aquat. Biol.* **2014**, *21*. [CrossRef]
11. Womble, J.N.; Gende, S.M. Post-Breeding Season Migrations of a Top Predator, the Harbor Seal (*Phoca vitulina richardii*), from a Marine Protected Area in Alaska. *PLoS ONE* **2013**, *8*, e55386. [CrossRef]
12. Reijnders, P.J.H.; Brasseur, S.M.J.M.; Meesters, E.H.W.G. Earlier pupping in harbour seals, *Phoca vitulina*. *Biol. Lett.* **2010**, *6*, 854–857. [CrossRef] [PubMed]
13. Robinson, K.J.; Hall, A.J.; Scholl, G.; Debier, C.; Thomé, J.P.; Eppe, G.; Adam, C.; Bennett, K.A. Investigating decadal changes in persistent organic pollutants in Scottish grey seal pups. *Aquat. Conserv. Mar. Freshw. Ecosyst.* **2019**, *29*, 86–100. [CrossRef]
14. Brasseur, S.M.; Reijnders, P.J.; Cremer, J.; Meesters, E.; Kirkwood, R.; Jensen, L.F.; Jeβ, A.; Galatius, A.; Teilmann, J.; Aarts, G. Echoes from the past: Regional variations in recovery within a harbour seal population. *PLoS ONE* **2018**, *13*, e0189674. [CrossRef] [PubMed]
15. Kovacs, K.M.; Lydersen, C. Climate change impacts on seals and whales in the North Atlantic Arctic and adjacent shelf seas. *Sci. Prog.* **2008**, *91*, 117–150. [CrossRef]
16. Forcada, J.; Trathan, P.N.; Reid, K.; Murphy, E.J. The effects of global climate variability in pup production of antarctic fur seals. *Ecology* **2005**, *86*, 2408–2417. [CrossRef]

17. Descamps, S.; Aars, J.; Fuglei, E.; Kovacs, K.M.; Lydersen, C.; Pavlova, O.; Pedersen, Å.Ø.; Ravolainen, V.; Strøm, H. Climate change impacts on wildlife in a High Arctic archipelago—Svalbard, Norway. *Glob. Chang. Biol.* **2017**, *23*, 490–502. [CrossRef]
18. Pörtner, H.-O.; Roberts, D.C.; Masson-Delmotte, V.; Zhai, P.; Tignor, M.; Poloczanska, E.; Mintenbeck, K.; Alegría, A.; Nicolai, M.; Okem, A.; et al. Summary for Policymakers. In IPCC Special Report on the Ocean and Cryosphere in a Changing Climate. Available online: https://www.ipcc.ch/srocc/chapter/summary-for-policymakers/ (accessed on 30 December 2020).
19. Parmesan, C.; Yohe, G. A globally coherent fingerprint of climate change impacts across natural systems. *Nature* **2003**, *421*, 37–42. [CrossRef]
20. Poloczanska, E.S.; Brown, C.J.; Sydeman, W.J.; Kiessling, W.; Schoeman, D.S.; Moore, P.J.; Brander, K.; Bruno, J.F.; Buckley, L.B.; Burrows, M.T.; et al. Global imprint of climate change on marine life. *Nat. Clim. Chang.* **2013**, *3*, 919–925. [CrossRef]
21. Kovacs, K.M.; Aguilar, A.; Aurioles, D.; Burkanov, V.; Campagna, C.; Gales, N.; Gelatt, T.; Goldsworthy, S.D.; Goodman, S.J.; Hofmeyr, G.J.; et al. Global threats to pinnipeds. *Mar. Mammal Sci.* **2012**, *28*, 414–436. [CrossRef]
22. Dippner, J.W.; Vuorinen, I.; Daunys, D.; Flinkman, J.; Halkka, A.; Köster, F.W.; Lehikoinen, E.; MacKenzie, B.R.; Möllmann, C.; Møhlenberg, F.; et al. Climate-related Marine Ecosystem Change. In *Assessment of Climate Change for the Baltic Sea Basin*; Springer: Berlin/Heidelberg, Germany, 2008; pp. 309–377.
23. Hamilton, C.D.; Kovacs, K.M.; Ims, R.A.; Aars, J.; Lydersen, C. An Arctic predator–prey system in flux: Climate change impacts on coastal space use by polar bears and ringed seals. *J. Anim. Ecol.* **1054**, *2017*, 86–1064. [CrossRef]
24. Hindell, M.A.; Sumner, M.; Bestley, S.; Wotherspoon, S.; Harcourt, R.G.; Lea, M.A.; Alderman, R.; McMahon, C.R. Decadal changes in habitat characteristics influence population trajectories of southern elephant seals. *Glob. Chang. Biol.* **2017**, *23*, 5136–5150. [CrossRef] [PubMed]
25. Costa, D.P.; Huckstadt, L.A.; Crocker, D.E.; McDonald, B.I.; Goebel, M.E.; Fedak, M.A. Approaches to Studying Climatic Change and its Role on the Habitat Selection of Antarctic Pinnipeds. *Integr. Comp. Biol.* **2010**, *1018*, 50–1030. [CrossRef] [PubMed]
26. Wilcox, C.; Hobday, A.J.; Chambers, L.E. Using expert elicitation to rank ecological indicators for detecting climate impacts on Australian seabirds and pinnipeds. *Ecol. Indic.* **2018**, *95*, 637–644. [CrossRef]
27. Rugh, D.; DeMaster, D.; Rooney, A.; Breiwick, J.; Shelden, K.; Moore, S. A review of bow-head whale (*Balaena mysticetus*) stock identity. *J. Cetacean Res. Manag.* **2003**, *7*, 1–12.
28. Berta, A.; Churchill, M. Pinniped taxonomy: Review of currently recognized species and subspecies, and evidence used for their description. *Mammals Rev.* **2012**, *42*, 207–234. [CrossRef]
29. Westlake, R.L.; O'Corry-Crowe, G.M. Macrogeographic Structure and Patterns of Genetic Diversity in Harbor Seals (*Phoca vitulina*) from Alaska to Japan. *J. Mammal.* **2002**, *1111*, 83–1126. [CrossRef]
30. DFO. Recovery Strategy for the Harbour Seal, Lacs Des Loups Marins Subspecies (*Phoca vitulina mellonae*). 2018. Available online: https://www.canada.ca/en/environment-climate-change/services/species-risk-public-registry/recovery-strategies/harbour-seal-lacs-loups-marins.html (accessed on 6 May 2019).
31. Stanley, H.F.; Casey, S.; Carnahan, J.M.; Goodman, S.; Harwood, J.; Wayne, R.K. Worldwide patterns of mitochondrial DNA differentiation in the harbor seal (*Phoca vitulina*). *Mol. Biol. Evol.* **1996**, *13*, 368–382. [CrossRef]
32. Olsen, M.T.; Andersen, L.W.; Dietz, R.; Teilmann, J. Integrating genetic data and population viability analyses for the identification of harbour seal (*Phoca vitulina*) populations and management units. *Mol. Ecol.* **2014**, 815–831. [CrossRef]
33. Andersen, L.; Olsen, M.T. Distribution and population structure of North Atlantic harbour seals (*Phoca vitulina*). *NAMMCO Sci. Publ.* **2010**, *8*, 15. [CrossRef]
34. Herreman, J.K.; Blundell, G.M.; Mcdonald, D.B.; Ben-David, M. Asymmetrical male-mediated gene flow between Harbor seal (*Phoca vitulina*) populations in Alaska. *Can. J. Zool.* **2009**, *87*, 498–507. [CrossRef]
35. Goodman, S.J. Patterns of extensive genetic differentiation and variation among European harbor seals (*Phoca vitulina vitulina*) revealed using microsatellite DNA poly morphism. *Mol. Biol. Evol.* **1998**, *15*, 104–118. [CrossRef] [PubMed]
36. O'Corry-Crowe, G.M.; Martien, K.K.; Taylor, B.L. The Analysis of Population Genetic Structure in Alaskan Harbor Seals, *Phoca vitulina*, as a Framework for the Identification of Management Stocks; Southwest Fisheries Science Center Administrative Report LJ-03-08. Available online: https://repository.library.noaa.gov/view/noaa/19225 (accessed on 16 December 2004).
37. Mizuno, M.; Kobayashi, M.; Sasaki, T.; Haneda, T.; Masubuchi, T. Current population genetics of Japanese harbor seals: Two distinct populations found within a small area. *Mar. Mammal Sci.* **2020**, *36*, 915–924. [CrossRef]
38. Steingass, S.; Horning, M.; Bishop, A.M. Space use of Pacific harbor seals (*Phoca vitulina* richardii) from two haul-out locations along the Oregon coast. *PLoS ONE* **2019**, *14*, e0219484. [CrossRef] [PubMed]
39. Mizuno, M.; Sasaki, T.; Kobayashi, M.; Haneda, T.; Masubuchi, T. Mitochondrial DNA reveals secondary contact in Japanese harbour seals, the southernmost population in the western Pacific. *PLoS ONE* **2018**, *13*, e0191329. [CrossRef] [PubMed]
40. Muto, M.M.; Helker, V.T.; Delean, B.J.; Angliss, R.P.; Boveng, P.L.; Breiwick, J.M.; Brost, B.M.; Cameron, M.F.; Clapham, P.J.; Dahle, S.P.; et al. Alaska Marine Mammal Stock Assessments. 2019. Available online: https://repository.library.noaa.gov/view/noaa/25642 (accessed on 31 July 2020).
41. Andersen, L.W.; Lydersen, C.; Frie, A.K.; Rosing-Asvid, A.; Hauksson, E.; Kovacs, K.M. A population on the edge: Genetic diversity and population structure of the world's northernmost harbour seals (*Phoca vitulina*). *Biol. J. Linn. Soc.* **2011**, *102*, 420–439. [CrossRef]
42. SCOS. Scientific Advice on Matters Related to the Management of Seal Populations. 2018. Available online: http://www.smru.st-andrews.ac.uk/files/2019/05/SCOS-2018.pdf (accessed on 31 May 2019).

43. Olsen, M.T.; Islas, V.; Graves, J.A.; Onoufriou, A.; Vincent, C.; Brasseur, S.; Frie, A.K.; Hall, A.J. Genetic population structure of harbour seals in the United Kingdom and neighbouring waters. *Aquat. Conserv. Mar. Freshw. Ecosyst.* **2017**, *27*, 839–845. [CrossRef]
44. Bjørge, A.; Desportes, G.; Waring, G.; Rosing-Asvid, A. Introduction: The harbour seal (*Phoca vitulina*)—A global perspective. In *NAMMCO Scientific Publications 8*; UiT The Arctic University of Norway: Tromsø, Norway, 2010; pp. 7–14.
45. Lowry, L.; Phoca vitulina. In the IUCN Red List of Threatened Species 2016. Available online: https://dx.doi.org/10.2305/IUCN.UK.2016-1.RLTS.T17013A45229114.en (accessed on 6 May 2019).
46. Kobayashi, Y.; Kariya, T.; Chishima, J.; Fujii, K.; Wada, K.; Baba, S.; Itoo, T.; Nakaoka, T.; Kawashima, M.; Saito, S.; et al. Population trends of the Kuril harbour seal *Phoca vitulina* stejnegeri from 1974 to 2010 in southeastern Hokkaido, Japan. *Endanger. Species Res.* **2014**, *24*, 61–72. [CrossRef]
47. Rosing-Asvid, A. Catch history and status of the harbour seal (*Phoca vitulina*) in Greenland. *NAMMCO Sci. Publ.* **2010**, *8*. [CrossRef]
48. Niizuma, A.; Hayama, S. A.; Hayama, S. A review of the taxonomy of the Kuril seal and other members of the genus Phoca (sensu stricto). In *Ecology and Protection of Kuril Seal*; Wada, K., Itoo, T., Niizum, A., Hayama, S., Suzuki, M., Eds.; Tokai University Press: Tokyo, Japan, 1986; pp. 1–18.
49. Trukhin, A. Current Status of Pinnnipeds in the Sea of Okhotsk. In Proceedings of the Fourth Workshop on the Okhotsk Sea and Adjacent Areas, Abashiri, Japan, 27–29 August 2008.
50. Mamaev, E.G. A new method of counting *Phoca vitulina* ssp. Stejnegeri (Phocida, Carnivora) on the Commander Islands (Russia). *Nat. Conserv. Res.* **2018**, *3*. [CrossRef]
51. Reijnders, P.J.H.; Brasseur, S.; Toorn, J.V.D.; Boyd, I.; Harwood, J.; Lowry, L. *Seals, Fur Seals, Sea Lions and Walruses: Status of Pinnipeds and Conservation Action Plan*; International Union for the Conservation of Nature and Natural Resources (IUCN): Gland, Switzerland, 1993.
52. Allen, B.; Angliss, R. Alaska Marine Mammal Stock Assessments. 2012. Available online: https://www.fisheries.noaa.gov/resource/document/alaska-marine-mammal-stock-assessments-2012 (accessed on 6 May 2019).
53. Jansen, J.K.; Boveng, P.L.; Hoef, J.M.V.; Dahle, S.P.; Bengtson, J.L. Natural and human effects on harbor seal abundance and spatial distribution in an Alaskan glacial fjord. *Mar. Mammal Sci.* **2015**, *31*, 66–89. [CrossRef]
54. Lydersen, C.; Assmy, P.; Falk-Petersen, S.; Kohler, J.; Kovacs, K.M.; Reigstad, M.; Steen, H.; Strøm, H.; Sundfjord, A.; Varpe, Ø.; et al. The importance of tidewater glaciers for marine mammals and seabirds in Svalbard, Norway. *J. Mar. Syst.* **2014**, *129*, 452–471. [CrossRef]
55. Boveng, P.L.; Hoef, J.M.V.; Withrow, D.E.; London, J.M. A Bayesian Analysis of Abundance, Trend, and Population Viability for Harbor Seals in Iliamna Lake, Alaska. *Risk Anal.* **1988**, *2018*, 38–2009. [CrossRef] [PubMed]
56. Burns, J.M.; Van Lanen, J.M.; Holen, D.L.; Zimpelman, G.; Jones, B.E.; Withrow, D.E.; Askoak, T.; Aderman, H.; O'Corey-Crowe, G. Integrating Local Traditional Knowledge and Subsistence Use Patterns with Aerial Surveys to Improve Scientific and Local Understanding of Iliamna Lake Seals. Alaska Department of Fish and Game, Division of Subsistence, 2016. Technical Paper No. 416. Anchorage. 2016. Available online: https://www.fisheries.noaa.gov/resource/peer-reviewed-research/integrating-local-traditional-knowledge-and-subsistence-use (accessed on 1 June 2016).
57. Hauser, D.D.W.; Allen, C.S.; Rich, H.B.; Quinn, T.P. Resident harbor seals (*Phoca vitulina*) in Iliamna Lake, Alaska: Summer diet and partial consumption of adult sockeye salmon (Oncorhynchus nerka). *Aquat. Mammals* **2008**, *34*, 303–309. [CrossRef]
58. Small, R.J.; Pendleton, G.W.; Pitcher, K.W. Trends in abundance of Alaska harbor seals, 1983–2001. *Mar. Mammal Sci.* **2003**, *19*, 44–362. [CrossRef]
59. Adkison, M.D.; Quinn, T.J.; Small, R.J. Evaluation of the Alaska harbor seal (*Phoca vitulina*) population survey: A simulation study. *Mar. Mammal Sci.* **2003**, *19*, 764–790. [CrossRef]
60. Small, R.J.; Boveng, P.L.; Byrd, G.V.; Withrow, D.E. Harbor seal population decline in the Aleutian Archipelago. *Mar. Mammal Sci.* **2008**, *24*, 845–863. [CrossRef]
61. Jemison, L.A.; Pendleton, G.W.; Wilson, C.A.; Small, R.J. Long-term trends in harbor seal numbers at Tugidak Island and Nanvak Bay, Alaska. *Mar. Mammal Sci.* **2006**, *22*, 339–360. [CrossRef]
62. Pitcher, K.W. Major Decline in Number of Harbor Seals, *Phoca vitulina richardsi*, on Tugidak Island, Gulf of Alaska. *Mar. Mammal Sci.* **1990**, *6*, 121–134. [CrossRef]
63. Frost, K.J.; Lowry, L.F.; Hoef, J.M.V. Monitoring the trend of harbor seals in Prince William Sound, Alaska, after the Exxon Valdez oil spill. *Mar. Mammal Sci.* **1999**, *15*, 494–506. [CrossRef]
64. Wang, D.; Atkinson, S.; Hoover-Miller, A.; Shelver, W.L.; Li, Q.X. Organic halogenated contaminants in mother–fetus pairs of harbor seals (*Phoca vitulina richardii*) from Alaska, 2000–2002. *J. Hazard. Mater.* **2012**, *224*, 72–78. [CrossRef]
65. Hoover-Miller, A.; Armato, P. Harbor seal use of glacier ice and terrestrial haul-outs in the Kenai Fjords, Alaska. *Mar. Mammal Sci.* **2018**, *34*, 616–644. [CrossRef]
66. Mathews, E.A.; Pendleton, G.W. Declines in harbor seal (*Phoca vitulina*) numbers in Glacier Bay National Park, Alaska, 1992-2002. *Mar. Mammal Sci.* **2006**, *22*, 167–189. [CrossRef]
67. Womble, J.N.; Pendleton, G.W.; Mathews, E.A.; Blundell, G.M.; Bool, N.M.; Gende, S.M. Harbor seal (*Phoca vitulina richardii*) decline continues in the rapidly changing landscape of Glacier Bay National Park, Alaska 1992–2008. *Mar. Mammal Sci.* **2010**, *26*, 686–697. [CrossRef]

68. Womble, J.N.; Hoef, J.M.V.; Gende, S.M.; Mathews, E.A. Calibrating and adjusting counts of harbor seals in a tidewater glacier fjord to estimate abundance and trends 1992 to 2017. *Ecosphere* **2020**, *11*, e03111. [CrossRef]
69. Miller, A.J.; Cayan, D.R.; Barnett, T.P.; Graham, N.E.; Oberhuber, J.M. The 1976-77 Climate Shift of the Pacific Ocean. *Oceanography* **1994**, *7*, 21–26. [CrossRef]
70. Trenberth, K.E.; Hurrell, J.W. Decadal atmosphere-ocean variations in the Pacific. *Clim. Dyn.* **1994**, *9*, 303–319. [CrossRef]
71. Trites, A.W.; Miller, A.J.; Maschner, H.D.; Alexander, M.A.; Bograd, S.J.; Calder, J.A.; Capotondi, A.; Coyle, K.O.; Lorenzo, E.D.; Finney, B.P.; et al. Bottom-up forcing and the decline of Steller sea lions (*Eumetopias jubatus*) in Alaska: Assessing the ocean climate hypothesis. *Fish. Oceanogr.* **2006**, *16*, 46–67. [CrossRef]
72. Bond, N.A.; Cronin, M.F.; Freeland, H.; Mantua, N. Causes and impacts of the 2014 warm anomaly in the NE Pacific. *Geophys. Res. Lett.* **2015**, *42*, 3414–3420. [CrossRef]
73. Batten, S.D.; Raitsos, D.E.; Danielson, S.; Hopcroft, R.; Coyle, K.; McQuatters-Gollop, A. Interannual variability in lower trophic levels on the Alaskan Shelf. *Deep Sea Res. Part II Top. Stud. Oceanogr.* **2018**, *147*, 58–68. [CrossRef]
74. Olesiuk, P.F. An Assessment of Population Trends and Abundance of Harbour Seals (*Phoca vitulina*) in British Columbia. *DFO Can. Sci. Advis. Sec. Res. Doc* **2009**, 105.
75. Harbour seal (*Phoca vitulina*) Counts and Haulout Locations in the Strait of Georgia, British Columbia Coast. Available online: https://open.canada.ca/data/en/dataset/be5a4ba8-79dd-4787-bf8a-0d460d25954c (accessed on 6 May 2019).
76. Huber, H.R.; Jeffries, S.J.; Brown, R.F.; Delong, R.L.; Vanblaricom, G. Correcting aerial survey counts of harbor seals (*Phoca vitulina richardsi*) in Washington and Oregon. *Mar. Mammal Sci.* **2001**, *17*, 276–293. [CrossRef]
77. Brown, R.F.; Wright, B.E.; Riemer, S.D.; Laake, J. Trends in abundance and current status of harbor seals in Oregon: 1977-2003. *Mar. Mammal Sci.* **2005**, *21*, 657–670. [CrossRef]
78. Sydeman, W.J.; Allen, S.G. Pinniped population dynamics in central California: Correlations with sea surface temperature and upwelling indices. *Mar. Mammal Sci.* **1999**, *15*, 446–461. [CrossRef]
79. Brown, R.F.; Jeffries, S.J.; Wright, B.E. Conductivity-Temperature-Depth Profiling of the Columbia River Mouth Using Pacific Harbor Seals as Sampling Platforms. 2013. Available online: https://apps.dtic.mil/dtic/tr/fulltext/u2/a598459.pdf (accessed on 6 May 2019).
80. Carretta, J.V.; Oleson, E.M.; Weller, D.W.; Lang, A.R.; Forney, K.A.; Baker, J.D.; Hanson, B.; Martien, K.K.; Muto, M.; Orr, A.J.; et al. U.S. Pacific Marine Mammal Stock Assessments, 2013. Available online: https://repository.library.noaa.gov/view/noaa/4772 (accessed on 4 January 2021).
81. Steingass, S.M. Habitat Use, Spatial Ecology, and Stable Isotope Variability of the Pacific Harbor Seal (*Phoca vitulina richardii*) along the Oregon Coast. Ph.D. Thesis, Oregon State University, Corvallis, OR, USA, 2018.
82. Becker, B.H.; Press, D.T.; Allen, S.G. Modeling the effects of El Niño, density-dependence, and disturbance on harbor seal (*Phoca vitulina*) counts in Drakes Estero, California: 1997–2007. *Mar. Mammal Sci.* **2009**, *25*, 1–18. [CrossRef]
83. Elorriaga-Verplancken, F.R.; Morales-Luna, L.; Heckel, G.; Schramm, Y. Foraging ecology of harbour seals (*Phoca vitulina*) and Northern elephant seals (Mirounga angustirostris) from Baja California, Mexico: Inferences from stable isotopes in pups. *J. Mar. Biol. Assoc. U. K.* **2016**, *96*, 903–908. [CrossRef]
84. Waring, G.T.; Gilbert, J.R.; Belden, D.; Van Atten, A.; DiGiovanni, R.A., Jr. A review of the status of harbour seals (*Phoca vitulina*) in the Northeast United States of America. *NAMMCO Sci. Publ.* **2010**, *8*, 191. [CrossRef]
85. Hammill, M.O.; Bowen, D.W.; Sjare, B. Status of harbour seals (*Phoca vitulina*) in Atlantic Canada. *NAMMCO Sci. Publ.* **2010**, *8*, 175–189. [CrossRef]
86. Boulva, J.; McLaren, I.A. Biology of the harbor seal, *Phoca vitulina*, in eastern Canada. *Bull. Fish. Res. Can.* **1979**, *24*. [CrossRef]
87. NOAA. Harbor Seal (*Phoca vitulina*): Western North Atlantic Stock. Available online: https://archive.afsc.noaa.gov/nmml/PDF/sars/ao2006sehr-wn.pdf (accessed on 6 May 2019).
88. Waring, G.T.; Digiovanni, R.A., Jr.; Josephson, E. 2012 Population Estimate for the Harbor Seal (*Phoca vitulina concolor*) in New England Waters. *NOAA Tech. Memo. NMFS NE* **2015**, *235*, 15. [CrossRef]
89. Lucas, Z.; Stobo, W.T. Shark-inflicted mortality on a population of harbour seals (*Phoca vitulina*) at Sable Island, Nova Scotia. *J. Zool.* **2000**, *252*, 405–414. [CrossRef]
90. Bowen, W.D.; Ellis, S.L.; Iverson, S.J.; Boness, D.J. Maternal and newborn life-history traits during periods of contrasting population trends: Implications for explaining the decline of harbour seals (*Phoca vitulina*), on Sable Island. *J. Zool.* **2003**, *261*, 155–163. [CrossRef]
91. Johnston, D.W.; Frungillo, J.; Smith, A.; Moore, K.; Sharp, B.; Schuh, J.; Read, A.J. Trends in Stranding and bycatch Rates of Gray and Harbor Seals along the Northeastern Coast of the United States: Evidence of Divergence in the Abundance of Two Sympatric Phocid Species? *PLoS ONE* **2015**, *10*, e0131660. [CrossRef] [PubMed]
92. COSEWIC. Status Appraisal Summary. 2018. Available online: https://www.canada.ca/en/environment-climate-change/services/species-risk-public-registry/cosewic-assessments-status-reports/harbour-seal-2018.html (accessed on 6 May 2019).
93. Teilmann, J.; Dietz, R. Status of the harbour seal, *Phoca vitulina*, in Greenland. *Oceanogr. Lit. Rev.* **1995**, *42*, 566.
94. Wiig, Ø. A description of common seals (*Phoca vitulina* L. 1758 from Svalbard. *Mar. Mammal Sci.* **1989**, *5*, 149–158. [CrossRef]
95. Lydersen, C.; Kovacs, K.M. Growth and population parameters of the world's northernmost harbour seals *Phoca vitulina* residing in Svalbard, Norway. *Polar Biol.* **2005**, *28*, 156–163. [CrossRef]

96. Henriksen, G.; Gjertz, I.; Kondakov, A. A review of the distribution and abundance of harbor seals, *Phoca vitulina*, on Svalbard, Norway, and in the Barents Sea. *Mar. Mammal Sci.* **1997**, *13*, 157–163. [CrossRef]
97. Blanchet, M.-A.; Lydersen, C.; Ims, R.A.; Kovacs, K.M. Making it through the first year: Ontogeny of movement and diving behavior in harbor seals from Svalbard, Norway. *Mar. Mammal Sci.* **2016**, *32*, 1340–1369. [CrossRef]
98. Blanchet, M.-A.; Lydersen, C.; Ims, R.A.; Kovacs, K.M. Seasonal, oceanographic and atmospheric drivers of diving behaviour in a temperate seal species living in the high arctic. *PLoS ONE* **2015**, *10*, e0132686. [CrossRef]
99. Hamilton, C.D.; Lydersen, C.; Ims, R.A.; Kovacs, K.M. Haul-Out Behaviour of the World's Northernmost Population of Harbour Seals (*Phoca vitulina*) throughout the Year. *PLoS ONE* **2014**, *4*, e86055. [CrossRef]
100. Leclerc, L.-M.E.; Lydersen, C.; Haug, T.; Bachmann, L.; Fisk, A.T.; Kovacs, K.M. A missing piece in the Arctic food web puzzle? Stomach contents of Greenland sharks sampled in Svalbard, Norway. *Polar Biol.* **1197**, *2012*, 35–1208. [CrossRef]
101. Routti, H.; Lydersen, C.; Hanssen, L.; Kovacs, K.M. Contaminant levels in the world's northernmost harbor seals (*Phoca vitulina*). *Mar. Pollut. Bull.* **2014**, 1–7. [CrossRef]
102. Hauksson, E.; Einarsson, S.T. Historical trend in harbour seal (*Phoca vitulina*) abundance in Iceland back to the year 1912. *NAMMCO Sci. Publ.* **2010**, *8*, 147. [CrossRef]
103. Granquist, S.M.; Hauksson, E. Seasonal, meteorological, tidal and diurnal effects on haul-out patterns of harbour seals (*Phoca vitulina*) in Iceland. *Polar Biol.* **2016**, *39*, 2347–2359. [CrossRef]
104. Hauksson, E. Sealing in Iceland in 1982–1989. *Hafrannsoknir* **1992**, *43*, 59–70.
105. Granquist, S.M.; Esparza-Salas, R.; Hauksson, E.; Karlsson, O.; Angerbjörn, A. Fish consumption of harbour seals (*Phoca vitulina*) in north western Iceland assessed by DNA metabarcoding and morphological analysis. *Polar Biol.* **2018**, *41*, 2199–2210. [CrossRef]
106. Punt, A.E.; Siple, M.; Sigurðsson, G.M.; Víkingsson, G.; Francis, T.B.; Granquist, S.M.; Hammond, P.S.; Heinemann, D.; Long, K.J.; Moore, J.E.; et al. Evaluating management strategies for marine mammal populations: An example for multiple species and multiple fishing sectors in Iceland. *Can. J. Fish. Aquat. Sci.* **2020**, *77*, 1316–1331. [CrossRef]
107. Þorbjornsson, J.G.; Hauksson, E.; Sigurðsson, G.M.; Granquist, S.M. Aerial Census of the Icelandic Harbour Seal Population in 2016: Population Estimate, Trends and Current Status. 2017. Available online: https://www.hafogvatn.is/is/midlun/utgafa/haf-og-vatnarannsoknir/aerial-census-of-the-icelandic-harbour-seal-phoca-vitulina-population-in-2016-population-estimate-trends-and-current-status-landselstalning-2016-stofnstaerdarmat-sveiflur-og-astand-stofns (accessed on 8 May 2019).
108. Granquist, S.M.; Hauksson, E. Population Estimate, Trends and Current Status of the Icelandic Harbour Seal (*Phoca vitulina*) Population in 2018. Marine and Freshwater Research in Iceland. HV 2019-36. 2019. Available online: https://www.hafogvatn.is/static/files/hv2019-36.pdf (accessed on 8 May 2019).
109. Bloch, D.; Mikkelsen, B.; Ofstad, L.H. Marine Mammals in Faroese Waters-With Special Attention to the South-South-Eastern Sector of the Region. 2000. Available online: http://projects.foib.fo/eia/Faroe_eia/Studies/Mammal_Final_Part1.pdf (accessed on 8 May 2019).
110. Mikkelsen, B. A note on the harbour seal (*Phoca vitulina*) in the Faroe Islands. *NAMMCO Sci. Publ.* **2010**, *8*, 143. [CrossRef]
111. Bjørge, A. Status of the harbour seal *Phoca vitulina* L. in Norway. *Biol. Conserv.* **1991**, *58*, 229–238. [CrossRef]
112. Hassani, S.; Dupuis, L.; Elder, J.F.; Caillot, E.; Gautier, G.; Hemon, A.; Lair, J.M.; Haelters, J. A note on harbour seal (*Phoca vitulina*) distribution and abundance in France and Belgium. *NAMMCO Sci. Publ.* **2010**, *8*, 107. [CrossRef]
113. Heide-Jørgensen, M.-P.; Härkönen, T.J. Rebuilding seal stocks in the Kattegat-Skagerrak. *Mar. Mammal Sci.* **1988**, *4*, 231–246. [CrossRef]
114. Nilssen, K.T.; Skavberg, N.-E.; Poltermann, M.; Haug, T.; Härkönen, T.; Henriksen, G. Status of harbour seals (*Phoca vitulina*) in mainland Norway. *NAMMCO Sci. Publ.* **2010**, *8*, 61. [CrossRef]
115. Härkönen, T.; Dietz, R.; Reijnders, P.; Teilmann, J.; Harding, K.; Hall, A.; Brasseur, S.; Siebert, U.; Goodman, S.J.; Jepson, P.D.; et al. The 1988 and 2002 phocine distemper virus epidemics in European harbour seals. *Dis. Aquat. Organ.* **2006**, *68*, 115–130. [CrossRef] [PubMed]
116. Zyryanov, S.V.; Egorov, S.A. Status of the harbour seal (*Phoca vitulina*) along the Murman coast of Russia. *NAMMCO Sci. Publ.* **2010**, *8*, 37. [CrossRef]
117. Mishin, V.L.; Zyryanov, S.V.; Goryaev, Y.I. Preliminary insight into the harbour seal population of the east Murmansk coast of the Barents Sea. *Mammalia* **2001**, *65*, 534–540. [CrossRef]
118. NAMMCO. Report of the NAMMCO Scientific Committee Meeting 4–7 November 2016. 2016. Available online: https://nammco.no/topics/scientific-committee-reports/ (accessed on 10 May 2019).
119. HELCOM. Population Trends and Abundance of Seals. HELCOM Core Indicator Report. 2018. Available online: https://www.helcom.fi/wp-content/uploads/2019/08/Population-trends-and-abundance-of-seals-HELCOM-core-indicator-2018.pdf (accessed on 10 May 2019).
120. Olsen, M.T.; Andersen, S.M.; Teilmann, J.; Dietz, R.; Edrén, S.M.C.; Linnet, A.; Härkönen, T. Status of the harbour seal (*Phoca vitulina*) in Southern Scandinavia. *NAMMCO Sci. Publ.* **2010**, *8*. [CrossRef]
121. Harkonen, T.; Andersen, S.M.; Teilmann, J.; Dietz, R.; Edrén, S.M.; Linnet, A.; Härkönen, T. Mass mortality in harbour seals and harbour porpoises caused by an unknown pathogen. *Vet. Rec.* **2008**, *162*, 555–556. [CrossRef]
122. Zohari, S.; Neimanis, A.; Härkönen, T.; Moraeus, C.; Valarcher, J.F. Avian influenza A(H10N7) virus involvement in mass mortality of harbour seals (*Phoca vitulina*) in Sweden, March through October 2014. *Euro Surveill.* **2014**, *19*. [CrossRef]

123. Olsson, M.; Karlsson, B.; Ahnland, E. Diseases and environmental contaminants in seals from the Baltic and the Swedish west coast. *Sci. Total Environ.* **1994**, *154*, 217–227. [CrossRef]
124. Härkönen, T.; Isakson, E. Status of harbour seals (*Phoca vitulina*) in the Baltic proper. *NAMMCO Sci. Publ.* **2010**, *8*, 71. [CrossRef]
125. Bjurlid, F.; Roos, A.; Jogsten, I.E.; Hagberg, J. Temporal trends of PBDD/Fs, PCDD/Fs, PBDEs and PCBs in ringed seals from the Baltic Sea (Pusa hispida botnica) between 1974 and 2015. *Sci. Total Environ.* **2018**, *616*, 1374–1383. [CrossRef]
126. Galatius, A.; Brasseur, S.M.J.M.; Busch, J.A.; Cremer, J.S.M.; Czeck, R.; Jeß, A.; Diederichs, B.; Körber, P.; Pund, R.; Siebert, U.; et al. Trilateral surveys of Harbour Seals in the Wadden Sea and Helgoland in 2019. Available online: https://www.waddensea-secretariat.org/resources/2019-harbour-seal-report (accessed on 8 May 2019).
127. Tougaard, J.; Henriksen, O.D.; Miller, L.A. Underwater noise from three types of offshore wind turbines: Estimation of impact zones for harbor porpoises and harbor seals. *J. Acoust. Soc. Am.* **2009**, *3766*, 125–3773. [CrossRef] [PubMed]
128. Vincent, C.; Huon, M.; Caurant, F.; Dabin, W.; Deniau, A.; Dixneuf, S.; Dupuis, L.; Elder, J.F.; Fremau, M.H.; Hassani, S.; et al. Grey and harbour seals in France: Distribution at sea, connectivity and trends in abundance at haul-out sites. *Deep Sea Res. Part II Top. Stud. Oceanogr.* **2017**, *141*, 294–305. [CrossRef]
129. Thompson, D.; Duck, C.D.; Morris, C.D.; Russell, D.J.F. The status of harbour seals (*Phoca vitulina*) in the UK. *Aquat. Conserv. Mar. Freshw. Ecosyst.* **2019**, *29*, 40–60. [CrossRef]
130. Thompson, D.; Duck, C.; Lonergan, M.E. The status of harbour seals (*Phoca vitulina*) in the United Kingdom. *NAMMCO Sci. Publ.* **2010**, *8*, 117. [CrossRef]
131. Lonergan, M.; Duck, C.; Moss, S.; Morris, C.; Thompson, D. Rescaling of aerial survey data with information from small numbers of telemetry tags to estimate the size of a declining harbour seal population. *Aquat. Conserv. Mar. Freshw. Ecosyst.* **2013**, *23*, 135–144. [CrossRef]
132. Lonergan, M.; Duck, C.D.; Thompson, D.; Mackey, B.L.; Cunningham, L.; Boyd, I.L. Using sparse survey data to investigate the declining abundance of British harbour seals. *J. Zool.* **2007**, *271*, 261–269. [CrossRef]
133. Wilson, L.J.; Hammond, P.S. The diet of harbour and grey seals around Britain: Examining the role of prey as a potential cause of harbour seal declines. *Aquat. Conserv. Mar. Freshw. Ecosyst.* **2019**, *29*, 71–85. [CrossRef]
134. Jones, E.L.; Sparling, C.E.; McConnell, B.J.; Morris, C.D.; Smout, S. Fine-scale harbour seal usage for informed marine spatial planning. *Sci. Rep.* **2017**, *7*, 11581. [CrossRef]
135. Boyd, I.L. Scientific Advice on Matters Related to the Management of Seal Populations: 2002. Natural Environment Research Council. Available online: http://www.smru.st-and.ac.uk/CurrentResearch.htm/scos.htm (accessed on 6 May 2019).
136. Jensen, S.K.; Lacaze, J.P.; Hermann, G.; Kershaw, J.; Brownlow, A.; Turner, A.; Hall, A. Detection and effects of harmful algal toxins in Scottish harbour seals and potential links to population decline. *Toxicon* **2015**, *97*, 1–14. [CrossRef]
137. Cronin, M.A. The status of the harbour seal (*Phoca vitulina*) in Ireland. *NAMMCO Sci. Publ.* **2010**, *8*. [CrossRef]
138. Kavanagh, A.S.; Cronin, M.A.; Walton, M.; Rogan, E. Diet of the harbour seal (*Phoca vitulina* vitulina) in the west and south-west of Ireland. *J. Mar. Biol. Assoc. U. K.* **2010**, *90*, 1517–1527. [CrossRef]
139. Cronin, M.; Duck, C.; Cadhla, O.Ó.; Nairn, R.; Strong, D.; O'Keeffe, C. An assessment of population size and distribution of harbour seals in the Republic of Ireland during the moult season in August 2003. *J. Zool.* **2007**, *273*, 131–139. [CrossRef]
140. Cronin, M.; Gregory, S.; Rogan, E. Moulting phenology of the harbour seal in south-west Ireland. *J. Mar. Biol. Assoc. U. K.* **2014**, *1079*, 94–1086. [CrossRef]
141. Luck, C.; Cronin, M.; Gosch, M.; Healy, K.; Cosgrove, R.; Tully, O.; Rogan, E.; Jessopp, M. Drivers of spatiotemporal variability in bycatch of a top marine predator: First evidence for the role of water turbidity in protected species bycatch. *J. Appl. Ecol.* **2020**, *57*, 219–228. [CrossRef]
142. Cosgrove, R.; Gosch, M.; Reid, D.; Sheridan, M.; Chopin, N.; Jessopp, M.; Cronin, M. Seal bycatch in gillnet and entangling net fisheries in Irish waters. *Fish. Res.* **2016**, *183*, 192–199. [CrossRef]
143. NOAA National Centers for Environmental Information. State of the Climate: Global Climate Report for August 2019. Available online: https://www.ncdc.noaa.gov/sotc/global/201908 (accessed on 31 December 2019).
144. Hansen, B.B.; Isaksen, K.; Benestad, R.E.; Kohler, J.; Larsen, J.O.; Varpe, Ø. Warmer and wetter winters: Characteristics and implications of an extreme weather event in the High Arctic. *Environ. Res. Lett.* **2014**, *9*, 114021. [CrossRef]
145. Simpkins, M.A.; Withrow, D.E.; Cesarone, J.C.; Boveng, P.L. Stability in the proportion of harbor seals hauled out under locally ideal conditions. *Mar. Mammal Sci.* **2003**, *19*, 791–805. [CrossRef]
146. Godsell, J. Herd formation and haul-out behaviour in harbour seals (*Phoca vitulina*). *J. Zool.* **1988**, *215*, 83–98. [CrossRef]
147. Hansen, S.; Lavigne, D.M. Ontogeny of the Thermal Limits in the Harbor Seal (*Phoca vitulina*) Ontogeny. *Physiol. Zool.* **1997**, *70*, 85–92. [CrossRef]
148. Comiso, J.C.; Hall, D.K. Climate trends in the Arctic as observed from space. *Wiley Interdiscip. Rev. Clim. Chang.* **2014**, *5*, 389–409. [CrossRef]
149. Maslanik, J.A.; Fowler, C.; Stroeve, J.; Drobot, S.; Zwally, J.; Yi, D.; Emery, W. A younger, thinner Arctic ice cover: Increased potential for rapid, extensive sea-ice loss. *Geophys. Res. Lett.* **2004**, *2007*, 34–2008. [CrossRef]
150. Bajzak, C.E.; Bernhardt, W.; Mosnier, A.; Hammill, M.O.; Stirling, I. Habitat use by harbour seals (*Phoca vitulina*) in a seasonally ice-covered region, the western Hudson Bay. *Polar Biol.* **2013**, *36*, 477–491. [CrossRef]
151. Lesage, V.; Hammill, M.O.; Kovacs, K.M. Long-distance movements of harbour seals (*Phoca vitulina*) from a seasonally ice-covered area, the St. Lawrence River estuary, Canada. *Can. J. Zool.* **2004**, *1070*, 82–1081. [CrossRef]

152. Florko, K.R.N. Decreasing sea ice conditions in western Hudson Bay and an increase in abundance of harbour seals (*Phoca vitulina*) in the Churchill River. *Polar Biol.* **2018**, *1187*, 41–1195. [CrossRef]
153. Baechler, J.; Beck, C.A.; Bowen, W.D. Dive shapes reveal temporal changes in the foraging behaviour of different age and sex classes of harbour seals (*Phoca vitulina*). *Can. J. Zool.* **2002**, *1569*, 80–1577. [CrossRef]
154. Ramasco, V.; Barraquand, F.; Biuw, M.; McConnell, B.; Nilssen, K.T. The intensity of horizontal and vertical search in a diving forager: The harbour seal. *Mov. Ecol.* **2015**, *3*, 1–16. [CrossRef]
155. Zemp, M.; Huss, M.; Thibert, E.; Eckert, N.; McNabb, R.; Huber, J.; Barandun, M.; Machguth, H.; Nussbaumer, S.U.; Gärtner-Roer, I.; et al. Global glacier mass changes and their contributions to sea-level rise from 1961 to 2016. *Nature* **2019**, *568*, 382–386. [CrossRef]
156. Wouters, B.; Gardner, A.S.; Moholdt, G. Global Glacier Mass Loss During the GRACE Satellite Mission (2002–2016). *Front. Earth Sci.* **2019**, *7*, 96. [CrossRef]
157. Fay, F.H. The role of ice in the ecology of marine mammals of the Bering Sea. *Oceanogr. Bering Sea* **1974**, *2*, 383–399.
158. Blundell, G.M.; Womble, J.N.; Pendleton, G.W.; Karpovich, S.A.; Gende, S.M.; Herreman, J.K. Use of glacial and terrestrial habitats by harbor seals in Glacier Bay, Alaska: Costs and benefits. *Mar. Ecol. Prog. Ser.* **2011**, *429*, 277–290. [CrossRef]
159. Lydersen, C.; Vaquie-Garcia, J.; Lydersen, E.; Christensen, G.N.; Kovacs, K.M. Novel terrestrial haul-out behaviour by ringed seals (Pusa hispida) in Svalbard, in association with harbour seals (*Phoca vitulina*). *Polar Res.* **2017**, *36*, 1374124. [CrossRef]
160. Cordes, L.S.; O'Corry-Crowe, G.; Small, R.J. Surreptitious sympatry: Exploring the ecological and genetic separation of two sibling species. *Ecol. Evol.* **2017**, *1725*, 7–1736. [CrossRef] [PubMed]
161. Trillmich, F.; Dellinger, T. The effects of El Niño on Galapagos pinnipeds. In *Pinnipeds and El Niño Ecological Studies (Analysis and Synthesis)*; Trillmich, F., Ono, K.A., Eds.; Springer: Berlin, Heidelberg, 1991; pp. 66–74.
162. Lusseau, D.; Williams, R.; Wilson, B.; Grellier, K.; Barton, T.R.; Hammond, P.S.; Thompson, P.M. Parallel influence of climate on the behaviour of Pacific killer whales and Atlantic bottlenose dolphins. *Ecol. Lett.* **2004**, *1068*, 7–1076. [CrossRef]
163. Friedlaender, A.S.; Johnston, D.W.; Halpin, P.N. Progress in Oceanography Effects of the North Atlantic Oscillation on sea ice breeding habitats of harp seals (Pagophilus groenlandicus) across the North Atlantic. *Prog. Oceanogr.* **2010**, *86*, 261–266. [CrossRef]
164. Trites, A.W.; Donnelly, C.P. The decline of Steller sea lions *Eumetopias jubatus* in Alaska: A review of the nutritional stress hypothesis. *Mammals Rev.* **2003**, *33*, 3–28. [CrossRef]
165. Steingass, S.; Horning, M. Individual-based energetic model suggests bottom up mechanisms for the impact of coastal hypoxia on Pacific harbor seal (*Phoca vitulina richardii*) foraging behavior. *J. Theor. Biol.* **2017**, *416*, 190–198. [CrossRef]
166. Steingass, S.M.; Naito, Y. Ocean Deoxygenation: Everyone's Problem. Causes, Impacts, Consequences and Solutions. 2020. Available online: https://portals.iucn.org/library/sites/library/files/documents/08.10DEOX.pdf (accessed on 30 December 2020).
167. Fossheim, M.; Primicerio, R.; Johannesen, E.; Ingvaldsen, R.B.; Aschan, M.M.; Dolgov, A.V. Recent warming leads to a rapid borealization of fish communities in the Arctic. *Nat. Clim. Chang.* **2015**, *5*, 673. [CrossRef]
168. Kortsch, S.; Primicerio, R.; Aschan, M.; Lind, S.; Dolgov, A.V.; Planque, B. Food-web structure varies along environmental gradients in a high-latitude marine ecosystem. *Ecography* **2019**, *42*, 295–308. [CrossRef]
169. Frainer, A.; Primicerio, R.; Kortsch, S.; Aune, M.; Dolgov, A.V.; Fossheim, M.; Aschan, M.M. Climate-driven changes in functional biogeography of Arctic marine fish communities. *Proc. Natl. Acad. Sci. USA* **2017**, *114*, 12202–12207. [CrossRef]
170. Randelhoff, A.; Reigstad, M.; Chierici, M.; Sundfjord, A.; Ivanov, V.; Cape, M.; Vernet, M.; Tremblay, J.É.; Bratbak, G.; Kristiansen, S. Seasonality of the Physical and Biogeochemical Hydrography in the Inflow to the Arctic Ocean Through Fram Strait. *Front. Marine Sci.* **2018**, *5*, 224. [CrossRef]
171. Ingvaldsen, R.B.; Gjøsæter, H.; Ona, E.; Michalsen, K. Atlantic cod (*Gadus morhua*) feeding over deep water in the high Arctic. *Polar Biol.* **2017**, *40*, 2105–2111. [CrossRef]
172. Andersen, S.M.; Lydersen, C.; Grahl-Nielsen, O.; Kovacs, K.M. Autumn diet of harbour seals (*Phoca vitulina*) at Prins Karls Forland, Svalbard, assessed via scat and fatty-acid analyses. *Can. J. Zool.* **2004**, *1230*, 82–1245. [CrossRef]
173. Colominas, R. Harbour Seal Diet in a Changing Arctic (Svalbard, Norway). Master's Thesis, University of Bergen, Bergen, Norway, 2012.
174. Thompson, P.M.; Mcconnell, B.J.; Tollit, D.J.; Mackay, A.; Hunter, C.; Racey, P.A. Comparative Distribution, Movements and Diet of Harbour and Grey Seals from Moray Firth, N.E. Scotland. *J. Appl. Ecol.* **1996**, *1572*, 33–1584. [CrossRef]
175. Thompson, P.M.; Tollit, D.J.; Corpe, H.M.; Reid, R.J.; Ross, H.M. Changes in haematological parameters in relation to prey switching in a wild population of harbour seals. *Funct. Ecol.* **1997**, *11*, 743–750. [CrossRef]
176. Weijerman, M.; Lindeboom, H.; Zuur, A.F. Regime shifts in marine ecosystems of the North Sea and Wadden Sea. *Mar. Ecol. Prog. Ser.* **2005**, *298*, 21–39. [CrossRef]
177. Rocha, J.C.; Peterson, G.; Bodin, Ö.; Levin, S. Cascading regime shifts within and across scales. *Science* **2018**, *362*, 1379–1383. [CrossRef] [PubMed]
178. Blanchet, M.-A.; Primicerio, R.; Frainer, A.; Kortsch, S.; Skern-Mauritzen, M.; Dolgov, A.V.; Aschan, M. The role of marine mammals in the Barents Sea foodweb. *ICES J. Mar. Sci.* **2019**, *76*, i37–i53. [CrossRef]
179. Ferguson, S.H.; Higdon, J.W.; Chmelnitsky, E.W. The Rise of Killer Whales as a Major Arctic Predator. In *A Little Less Arctic*; Ferguson, S.H., Loseto, L.L., Mallory, M.L., Eds.; Springer: Dordrecht, The Netherlands, 2012.

180. Higdon, J.W.; Hauser, D.D.W.; Ferguson, S.H. Killer whales (*Orcinus orca*) in the Canadian Arctic: Distribution, prey items, group sizes, and seasonality. *Mar. Mammal Sci.* **2012**, *28*, E93–E109. [CrossRef]
181. Tryland, M.; Godfroid, J.; Arneberg, P. Impact of Climate Change on Infectious Diseases of Animals in the Norwegian Arctic. 2009. Norwegian Polar Institute Brief Report Series. Available online: https://brage.npolar.no/npolar-xmlui/handle/11250/172981 (accessed on 30 December 2020).
182. VanWormer, E.; Mazet, J.A.; Hall, A.; Gill, V.A.; Boveng, P.L.; London, J.M.; Gelatt, T.; Fadely, B.S.; Lander, M.E.; Sterling, J.; et al. Viral emergence in marine mammals in the North Pacific may be linked to Arctic sea ice reduction. *Sci. Rep.* **2019**, *9*, 15569. [CrossRef]
183. Polley, L.; Thompson, R.C.A. Parasite zoonoses and climate change: Molecular tools for tracking shifting boundaries. *Trends Parasitol.* **2009**, *25*, 285–291. [CrossRef]
184. Brooks, D.R.; Hoberg, E.P. How will global climate change affect parasite-host assemblages? *Trends Parasitol.* **2007**, *23*, 571–574. [CrossRef] [PubMed]
185. Bradley, M.J.; Kutz, S.J.; Jenkins, E.; O'Hara, T.M. The potential impact of climate change on infectious diseases of Arctic fauna. *Int. J. Circumpolar Health* **2005**, *64*, 468–477. [CrossRef] [PubMed]
186. Keroack, C.D.; Williams, K.M.; Fessler, M.K.; DeAngelis, K.E.; Tsekitsidou, E.; Tozloski, J.M.; Williams, S.A. A novel quantitative real-time PCR diagnostic assay for seal heartworm (Acanthocheilonema spirocauda) provides evidence for possible infection in the grey seal (Halichoerus grypus). *Int. J. Parasitol. Parasites Wildl.* **2018**, *7*, 147–154. [CrossRef] [PubMed]
187. Lehnert, K.; Schwanke, E.; Hahn, K.; Wohlsein, P.; Siebert, U. Heartworm (Acanthocheilonema spirocauda) and seal louse (Echinophthirius horridus) infections in harbour seals (*Phoca vitulina*) from the North and Baltic Seas. *J. Sea Res.* **2016**, *113*, 65–72. [CrossRef]
188. Burek, K.A.; Gulland, F.M.D.; O'Hara, T.M. Effects of climate change on arctic marine mammal health. *Ecol. Appl.* **2008**, *18*, S126–S134. [CrossRef]
189. Galaktionov, K.V. Patterns and processes influencing helminth parasites of Arctic coastal communities during climate change. *J. Helminthol.* **2017**, *91*, 387–408. [CrossRef]
190. Harvell, C.D.; Kim, K.; Burkholder, J.M.; Colwell, R.R.; Epstein, P.R.; Grimes, D.J.; Hofmann, E.E.; Lipp, E.K.; Osterhaus, A.D.; Overstreet, R.M.; et al. Emerging marine diseases—Climate links and anthropogenic factors. *Science* **1999**, *1505*, 285–1510. [CrossRef]
191. Marcogliese, D.J. Implications of climate change for parasitism of animals in the aquatic environment. *Can. J. Zool.* **2001**, *79*, 1331–1352. [CrossRef]
192. Siebert, U.; Gulland, F.; Harder, T.; Jauniaux, T.; Seibel, H.; Wohlsein, P.; Baumgärtner, W. Epizootics in harbour seals (*Phoca vitulina*): Clinical aspects. *NAMMCO Sci. Publ.* **2010**, *8*, 265. [CrossRef]
193. Stokholm, I.; Härkönen, T.; Harding, K.C.; Siebert, U.; Lehnert, K.; Dietz, R.; Teilmann, J.; Galatius, A.; Havmøller, L.W.; Carroll, E.L.; et al. Phylogenomic insights to the origin and spread of phocine distemper virus in European harbour seals in 1988 and 2002. *Dis. Aquat. Organ.* **2019**, *133*, 47–56. [CrossRef]
194. Lavigne, D.M.; Schmitz, O.J. Global warming and increasing population densities: A prescription for seal plagues. *Mar. Pollut. Bull.* **1990**, *21*, 280–284. [CrossRef]
195. Rosing-Asvid, A.; Teilmann, J.; Dietz, R.; Olsen, M. First Confirmed Record of Grey Seals in Greenland. *Arctic* **2010**, *63*, 471–473. [CrossRef]
196. Sommer, S. The importance of immune gene variability (MHC) in evolutionary ecology and conservation. *Front. Zool.* **2005**, *2*, 16. [CrossRef] [PubMed]
197. Fujii, K.; Kobayashi, M.; Kai, C. Seroepidemiological survey of morbillivirus infection in Kuril harbor seals *Phoca vitulina* stejnegeri of Hokkaido, Japan. *Jpn. J. Vet. Res.* **2006**, *54*, 109–117. [PubMed]
198. Goldstein, T.; Mazet, J.A.K.; Gill, V.A.; Doroff, A.M.; Burek, K.A.; Hammond, J.A. Phocine distemper virus in northern sea otters in the Pacific Ocean, Alaska, USA. *Emerg. Infect. Dis.* **2009**, *15*, 925–927. [CrossRef]
199. Sanderson, C.E.; Alexander, K.A. Unchartered waters: Climate change likely to intensify infectious disease outbreaks causing mass mortality events in marine mammals. *Glob. Chang. Biol.* **2020**, *26*, 4284–4301. [CrossRef]
200. Jensen, S.K.; Aars, J.; Lydersen, C.; Kovacs, K.M.; Åsbakk, K. The prevalence of Toxoplasma gondii in polar bears and their marine mammal prey: Evidence for a marine transmission pathway? *Polar Biol.* **2010**, *33*, 599–606. [CrossRef]
201. Gibson, A.K.; Raverty, S.; Lambourn, D.M.; Huggins, J.; Magargal, S.L.; Grigg, M.E. Polyparasitism is associated with increased disease severity in Toxoplasma gondii-infected marine sentinel species. *PLoS Negl. Trop. Dis.* **2011**, *5*, e1142. [CrossRef]
202. Corbett, J.J.; Lack, D.A.; Winebrake, J.J.; Harder, S.; Silberman, J.A.; Gold, M. Arctic shipping emissions inventories and future scenarios. *Atmos. Chem. Phys.* **2010**, *10*, 9689–9704. [CrossRef]
203. Jones, E.L.; Hastie, G.D.; Smout, S.; Onoufriou, J.; Merchant, N.D.; Brookes, K.L.; Thompson, D. Seals and shipping: Quantifying population risk and individual exposure to vessel noise. *J. Appl. Ecol.* **2017**, *54*, 1930–1940. [CrossRef]
204. Ahrens, L.; Siebert, U.; Ebinghaus, R. Temporal trends of polyfluoroalkyl compounds in harbor seals (*Phoca vitulina*) from the German Bight, 1999–2008. *Chemosphere* **2009**, *76*, 151–158. [CrossRef] [PubMed]
205. Blévin, P.; Aars, J.; Andersen, M.; Blanchet, M.A.; Hanssen, L.; Herzke, D.; Jeffreys, R.M.; Nordøy, E.S.; Pinzone, M.; de la Vega, C.; et al. Pelagic vs Coastal—Key Drivers of Pollutant Levels in Barents Sea Polar Bears with Contrasted Space-Use Strategies. *Environ. Sci. Technol.* **2020**, *54*, 985–995. [CrossRef]

206. Mos, L.; Morsey, B.; Jeffries, S. Chemical and biological pollution contribute to the immunological profiles of free-ranging harbor seals. *Env. Toxicol Chem* **2006**, *25*, 110–117. [CrossRef] [PubMed]
207. Edwards, M.; Johns, D.G.; Leterme, S.C.; Svendsen, E.; Richardson, A.J. Regional climate change and harmful algal blooms in the northeast Atlantic. *Limnol. Oceanogr.* **2006**, *51*, 820–829. [CrossRef]
208. Harvey, J.T. *Phoca vitulina* ssp. richardii. The IUCN Red List of Threatened Species 2016. Available online: https://dx.doi.org/10.2305/IUCN.UK.2016-1.RLTS.T17022A66991556.en (accessed on 17 September 2020).
209. Temple, H.J.; Terry, A. The Status and Distribution of European Mammals. 2007. Available online: https://ec.europa.eu/environment/nature/conservation/species/redlist/downloads/European_mammals.pdf (accessed on 17 September 2020).
210. Matsuda, H.; Yamamura, O.; Kitakado, T.; Kobayashi, Y.; Kobayashi, M.; Hattori, K.; Kato, H. Beyond dichotomy in the protection and management of marine mammals in Japan. *Therya* **2015**, *6*, 283–296. [CrossRef]

oceans

MDPI

Article

Antibiotic Resistance of Bacteria in Two Marine Mammal Species, Harbor Seals and Harbor Porpoises, Living in an Urban Marine Ecosystem, the Salish Sea, Washington State, USA

Stephanie A. Norman [1,2,3,*], Dyanna M. Lambourn [4], Jessica L. Huggins [5], Joseph K. Gaydos [6], Sandra Dubpernell [2], Susan Berta [2], Jennifer K. Olson [7], Victoria Souze [8], Alysha Evans [8], Betsy Carlson [9], Mandi Johnson [9], Rachel Mayer [10], Cathy King [3] and Alyssa Scott [7]

1 Marine-Med: Marine Research, Epidemiology, and Veterinary Medicine, Bothell, WA 98021, USA
2 Orca Network/Central Puget Sound Marine Mammal Stranding Network, Freeland, WA 98249, USA; dubpernells@yahoo.com (S.D.); susan@orcanetwork.org (S.B.)
3 World Vets, Gig Harbor, WA 98332, USA; drcathy@worldvets.org
4 Marine Mammal Investigations, Washington Department of Fish and Wildlife, Lakewood, WA 98498, USA; Dyanna.Lambourn@dfw.wa.gov
5 Cascadia Research Collective, Olympia, WA 98501, USA; JHuggins@cascadiaresearch.org
6 SeaDoc Society, UC Davis Wildlife Health Center—Orcas Island Office, Eastsound, WA 98245, USA; jkgaydos@ucdavis.edu
7 The Whale Museum, Friday Harbor, WA 98250, USA; jolson296@gmail.com (J.K.O.); alyssa@whalemuseum.org (A.S.)
8 Whatcom County Marine Mammal Stranding Network, Lummi Island, WA 98262, USA; vsouze56@gmail.com (V.S.); adelsby@gmail.com (A.E.)
9 Port Townsend Marine Science Center, Port Townsend, WA 98368, USA; bcarlson@ptmsc.org (B.C.); mjohnson@ptmsc.org (M.J.)
10 Sno-King Marine Mammal Response, Seattle, WA 98117, USA; Rachel.Mayer@comcast.net
* Correspondence: stephanie@marine-med.com

Citation: Norman, S.A.; Lambourn, D.M.; Huggins, J.L.; Gaydos, J.K.; Dubpernell, S.; Berta, S.; Olson, J.K.; Souze, V.; Evans, A.; Carlson, B.; et al. Antibiotic Resistance of Bacteria in Two Marine Mammal Species, Harbor Seals and Harbor Porpoises, Living in an Urban Marine Ecosystem, the Salish Sea, Washington State, USA. *Oceans* **2021**, *2*, 86–104. https://doi.org/10.3390/oceans2010006

Academic Editor: Alexander Werth
Received: 4 November 2020
Accepted: 19 January 2021
Published: 25 January 2021

Abstract: The pervasive use of antibiotics in human medicine, veterinary medicine, and agriculture can result in a significant increase in the spread and environmental persistence of antibiotic resistance in marine ecosystems. This study describes the presence and distribution of antibiotic-resistant bacteria in Salish Sea harbor seals (*Phoca vitulina*) and harbor porpoises (*Phocoena phocoena*) and evaluates species, age class, and geographic differences in resistance patterns. Isolates from 95 dead-stranded animals (74 seals/21 porpoises) were tested for resistance to a suite of 15 antibiotics. Of the 95 sampled, 85 (89%) (67 seals/18 porpoises) successfully yielded 144 isolates, with 37% resistant to at least one antibiotic and 26% multi-drug resistant (24% and 39% of seal and porpoise isolates, respectively). Overall, and by study region, porpoises were significantly more likely to harbor resistant organisms compared to seals. Significant differences between age classes were noted for the antibiotics amoxicillin, cephalexin, and cefovecin. Overall isolate resistance was significantly greater in porpoises than seals for several individual antibiotics. Multiple antibiotic resistance (MAR) indices greater than 0.2 were observed in 55% of multi-drug resistant isolates, suggesting seal and porpoise exposure to anthropogenic pollution. The relatively high and disparate prevalence of antibiotic resistance in these common, but ecologically dissimilar, marine mammals reflects a potentially large environmental pool of antibiotic resistant organisms in the Salish Sea or inherently different resistance gene patterns between the two species.

Keywords: antibiotic resistance; antimicrobial; multi-drug resistance; MAR index; harbor porpoise; harbor seal; marine ecosystem; Salish Sea

1. Introduction

Normal bacterial flora of an animal can shape its growth, development, and behavior, as well as mate selection [1]. However, host flora may change when antibiotic-resistant microbes are introduced into the organism [1]. Antibiotic resistance is a global concern [2] and has been characterized as the "quintessential One Health issue" [3,4]. The ubiquitous use of antibiotics in human and veterinary disease treatment and agriculture has resulted in a significant increase in the spread and environmental persistence of antibiotic resistance [5,6]. This includes the release of wastes that carry both antibiotics and antibiotic-resistant bacteria into the coastal marine environment [7]. Antibiotic-resistant microbes and genes within the aquatic environment have been documented in various marine species from cephalopods to marine mammals and elasmobranchs, such as sharks [6,8–13].

Although antibiotic-resistant bacteria have been documented in multiple marine species, most wild animals have never been directly exposed to antibiotics. When wastewater discharge, carrying antibiotics and resistant bacteria into terrestrial waterways, finds its way to marine coastlines, it may cause disease in marine organisms, contributing to antibiotic resistance [14]. This, and other anthropogenic contributions, may elevate natural background levels of antibiotic resistance genes in aquatic environments, encouraging their transfer into pathogens or serving as a means for antibiotic resistance propagation [15]. When resistant bacteria are introduced to animals or their environment, the animals may become sick, or resistance traits may be transferred to other bacterial species, or they may become a reservoir that transfers the bacteria and resistance back to humans and the environment [16].

Information on antibiotic resistance in marine species in Washington State's inland marine waters, collectively referred to as the Salish Sea, is relatively limited. Preliminary work reported resistance in young stranded harbor seals (*Phoca vitulina*) in rehabilitation [17], harbor seal pups found dead during beach searches [18], and in local endangered southern resident killer whales (SRKW) (*Orcinus orca*) breath and feces [19,20]. To determine the types and degree of antibiotic resistance in local marine mammal populations and to begin to better understand how resistance moves through wild animals and ecosystems, further examination is needed in this region on a wider-breadth of species, age classes, and locations throughout the Salish Sea. Additionally, investigating marine mammal species that are sympatric with resident killer whales, especially within this urban marine ecosystem, would also give insight into the ways in which antibiotic-resistant organisms potentially threaten the health of this endangered population [19,20]. The objective of this study was to evaluate the presence and distribution of antibiotic-resistant bacteria in Salish Sea harbor seals and harbor porpoises (*Phocoena phocoena*), with specific goals to describe the presence of antibiotic-resistant bacteria, determine differences in resistance between and within the two species, and describe geographic patterns.

2. Materials and Methods

2.1. Study Population

A cross-sectional, opportunistic, sampling study was conducted from October 2018–May 2020 to determine the prevalence (%) of antibiotic-resistant bacteria from samples of fresh dead (Code 2) [21], stranded marine mammals in the inland waters of Washington State (Salish Sea). Specifically, two local species were targeted for sampling: harbor seals and harbor porpoises. These species were chosen for several reasons: they both occur in the Salish Sea in the greatest numbers compared to other marine mammal species; they are the most commonly stranded species within the Sea [22,23]; harbor seal and porpoise populations inhabiting the Salish Sea generally tend to stay more localized without traveling great distances, compared to their outer coast cohorts [24]; and these two species would be the most likely to carry antibiotic-resistant bacteria originating from terrestrial sources surrounding the Salish Sea, compared to more migratory or less common marine mammal species within the Sea. Lastly, they provide an opportunity to compare a completely aquatic species (porpoise) to one that is semi-aquatic (seals). Age class determination (adult, subadult/juvenile, pup/calf) for harbor seals was based on size and time of year [25] and on straight length for porpoises using ranges from known-age animals examined in Washington State [26].

2.2. Study Area

Two general sampling regions within Washington State's portion of the Salish Sea were investigated: the northern region encompassing the San Juan Island archipelago and waters north to the Canadian border, south to the southern tip of Whidbey Island, and west to the entrance of the Strait of Juan de Fuca; and the southern region, starting just south of Whidbey Island and encompassing Hood Canal, the Seattle/Tacoma Metropolitan, and continuing south to Olympia (Figure 1). The sampling regions represented areas with variable anthropogenic activities and human population density. Tourists heavily visit the northern portion during the summer months, but less so the rest of the year; however, currents may transport untreated wastewaters down from the cities of Vancouver and Victoria, British Columbia into this region [27]. The lower half of the northern region contains two military bases (Naval Air Station Whidbey Island and Naval Station Everett), heavily visited Whidbey and Camano Islands, as well as agricultural areas on these islands and the mainland at corresponding latitudes. The southern sampling region encompasses major areas of dense human population, Seattle and Tacoma, as well as many passages, inlets, and islands, with varying degrees of industrial development, agriculture, or population density.

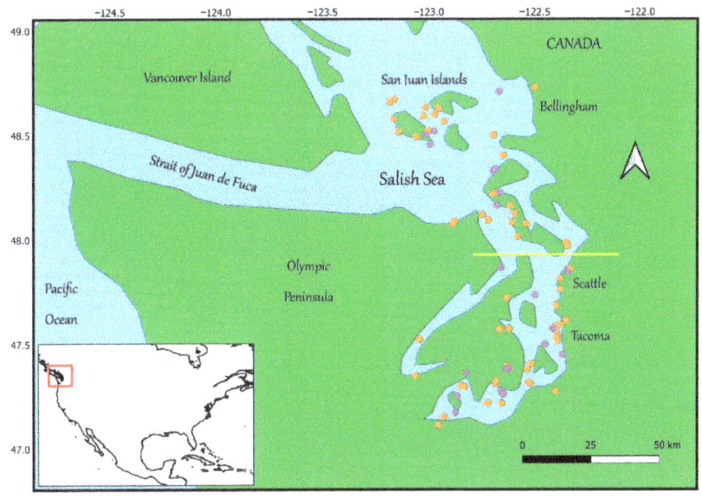

Figure 1. Distribution of stranded harbor seals (orange dots) and harbor porpoises (purple dots) sampled for antibiotic resistant organisms in the Salish Sea, Washington State, USA. Yellow horizontal line delineates northern and southern portions of the study area.

2.3. Sample Collection

All animals were externally examined and, if feasible, a necropsy (or thorough internal examination) was performed within 24 h of being recovered from the beach (stored at 4 °C overnight if not done the same day). If a necropsy could not be performed during this time frame, a swab was collected rectally to sample the large intestine, and the carcass was frozen for examination at a later date or left at the stranding site. Samples for aerobic culture and sensitivity were collected postmortem from the large intestine/feces of each animal either by careful placement of a sterile swab rectally to prevent skin or fur contamination of the swab, or during examination of the large intestine at necropsy. Additional swabs were collected from any lesions noted on gross external and/or internal examination that were suspicious for bacterial involvement such as swelling, discharge, or redness [17]. Sampling of wounds or lesions was conducted using standard techniques to minimize environmental contamination [28]. All swabs were held in Amies transport medium without charcoal (BBL™ CultureSwab™ Plus Collection and Transport Swabs, Beckton Dickinson and Co.,

Sparks, MD, USA). Samples were refrigerated and shipped overnight to a local veterinary reference laboratory for processing (Phoenix Laboratory, Mukilteo, WA, USA).

2.4. Bacterial Isolation and Antimicrobial Sensitivity Testing

Targeted bacteria included those that are of interest in marine mammal health, and overlap with health of terrestrial animals, humans, and other marine animals such as fish. They included *Streptococcus* spp. (alpha, beta, and gamma), *Staphylococcus aureus*, *Vibrio* spp., *Aeromonas hydrophila*, *Proteus vulgaris*, *Klebsiella* spp. (respiratory samples), and *Clostridium perfringens*. Given the extensive microbiome of the gastrointestinal tract, select organisms commonly considered enteric pathogens were targeted, and included *Salmonella*, *Clostridium* spp., *Campylobacter*, and *Escherichia coli*.

For each bacterial isolate, the following data were collected: animal and tissue of origin, stranding location coordinates of animal from which sample(s) were collected, taxonomic identification of isolate by the public veterinary laboratory, and sensitivity to each of the antibiotics tested. Standard methods were used to identify the bacteria, including growth on appropriate selective and differential media, gross colony appearance, morphology on gram stain, and biochemical reaction.

Conditions for aerobic bacterial culture were implemented, and bacterial isolates were identified based on colony morphology, growth characteristics, gram stain, and biochemical testing. For swabs of the large intestine (rectally or from descending colon at necropsy), a battery of primary plating media was used for routine bacterial fecal culture. Routine fecal culture setup was designed to optimize the recovery of *Salmonella*, *Shigella*, *Campylobacter*, and *E. coli*; thus, fecal specimens received for culture were plated onto at least four media: (i) MacConkey (MAC) agar, (ii) tryptic soy agar with 5% sheep blood, (iii) a selective/differential medium designed for the recovery of *Salmonella* and *Shigella*, and (iv) a medium designed for the recovery of *Campylobacter* [29]. In addition, blood agar plates were used to aid with the recovery of *Aeromonas* spp., *Plesiomonas* spp., and *Vibrio* spp., while Hektoen enteric agar was used to isolate and differentiate members of the species *Salmonella* and *Shigella*. Phenyl ethyl alcohol agar with 5% sheep blood was used to cultivate gram positive bacteria. The inoculated media was placed in a 37 °C incubator overnight. The following day, isolates were sub-cultured to isolate organisms. Identification of individual bacterial species was based on pH change indicators for substrate utilization, production of metabolic byproducts, and hydrolysis of substrates.

Swabs from tissues other than feces were streaked onto tryptic soy agar with 5% sheep blood, chocolate agar, Columbia CNA agar with 5% sheep blood, MacConkey agar, and thioglycolate agar as indicated for bacterial isolation. The inoculated media was placed in a 37 °C incubator overnight. The following day, isolates were subcultured to isolate out the targeted pathogenic organisms, with the number of isolates retrieved from a plate varying depending on the type of bacteria. Individual profiles of antimicrobial susceptibility testing and identification were developed using the same protocol, equipment, and antibiotic panel applied for the fecal cultures. Bacteria considered indicative of contamination, or those that were considered nonpathogenic, were presumptively identified and did not include susceptibility testing. A microbe was considered multi-drug resistant if an isolate was not considered susceptible to at least two or more bactericidal or bacteriostatic agents in at least two antimicrobial classes (adapted from Sweeney et al., 2018 [30]). *E. coli* isolates obtained from this study were archived by adding glycerol to a final concentration of 10–20%, freezing the culture, and storing it in an ultra-low-temperature freezer at −80 °C.

Further identification of Gram-negative bacteria was conducted using an Analytical Profile Index (API) 20E system (Biomérieux, Marcy-l'Étoile, France). Briefly, the bacterial suspensions were inoculated on a strip of 20 dehydrated reagents. These reagents are miniature versions of traditionally utilized biochemical assays for identifying bacterial species. The strip panel was incubated from 24 to 48 h at 35 °C. A numerical value, the API code, matched to the corresponding bacterial species in the API database, was assigned to the results. Bacterial isolates were not further confirmed through molecular testing.

Microaerophilic species such as *Campylobacter* were isolated using enriched media such as Skirrow agar and were grown in generator envelopes delivering 6% oxygen, 10% carbon dioxide, and 84% nitrogen for up to five days at 42 °C [29].

Individual profiles of antimicrobial susceptibility testing and identification were developed for each bacterial isolate of interest using an automated VITEK 2 instrument (Biomérieux, Marcy-l'Étoile, France). The antibiotic panel is a standard suite used by a local veterinary reference laboratory (Phoenix Laboratory, Mukilteo, WA, USA) and represents many of the most commonly used antibiotic classes in veterinary (and human) medicine in the United States. Antibiotics tested against isolates of clinical relevance included aminoglycosides (amikacin, gentamicin), carbapenims (imipenem), cephalosporins (cephalexin, cefovecin, cefpodoxime, ceftiofur), fluoroquinolones (enrofloxacin, marbofloxacin), penicillins (amoxicillin, amoxicillin/clavulanic acid), (sulfonamides) trimethoprim-sulfamethoxazole, tetracyclines (doxycycline), chloramphenicol, and florfenicol. The same set of antibiotics was used for a given species of bacteria, regardless of the location or timing of sampling, with results expressed as sensitive, intermediate, or resistant. For this study, isolates determined to be intermediate were not included in calculating resistance. Bacteria considered indicative of contamination or nonpathogenic were presumptively identified and were not subject to susceptibility testing.

2.5. Analysis

Descriptive frequencies were calculated for each isolated bacterium from both collection regions (North and South Salish Sea) and for both species, including prevalence of single and multi-drug antibiotic resistance and occurrence of antibiotic resistance within taxonomic groups of bacterial isolates. Chi-square tests were used to compare resistance to each class of antibiotics by bacterial isolate and to analyze resistance patterns between the porpoises and seals and by age class. To avoid the chance of false discovery rate due to multiple simultaneous comparisons of antibiotics, a separate chi-square test was run for each drug. The proportions of resistance for each isolate and for all isolates pooled were compared between the two sampling regions using logistic regression. Odds ratios (OR) with their 95% confidence intervals (CI) were calculated to estimate risk of antibiotic resistance between the two species, two regions, and three age classes. For cells with expected counts of <5, Fisher's exact test was used. To analyze differences in the level of antibiotic resistance (defined as the mean number (count) of antibiotics to which a bacterial isolate was resistant) between seals and porpoises, we applied a generalized linear model with a negative binomial distribution to account for any over-dispersion of the data. Statistical significance was considered at p-value < 0.05. Analyses were performed using STATA 15.0 (STATA, College Station, TX, USA).

The proportion of antibiotics to which a particular isolate was resistant was used to generate a Multiple Antibiotic Resistance Index (MAR: range 0 to 1), which has been used to reflect potential anthropogenic impacts and degree of antibiotic exposure for an environmental isolate [31]. The index was calculated as the ratio of the number of resistant antibiotics to which an isolate is resistant to the total number of antibiotics to which it was tested. The resulting indices were further grouped based on whether their value was 0, $\leq$0.2 (amount of antibiotic resistance typical of nonpoint sources of anthropogenic pollution) and >0.2 (amount of antibiotic resistance considered characteristic of point-source pollution) [31–33]. Due to the inherent multi-drug resistance of many *Pseudomonas* spp. isolates, an additional calculation of MAR indices, without inclusion of this genus, was conducted. Water sources with a MAR index > 0.4 are usually from human fecal origin and those <0.4 from nonhuman fecal contamination [34,35]. Lastly, spatial patterns of multi-drug resistance were determined by looking for clusters in the data using the program SaTScan. A Bernoulli model for spatial clusters was used, limiting cluster radius to 5 km and significant if p-value < 0.05 [36].

3. Results

A total of 95 animals were sampled (74 harbor seals [40 female:34 male], 21 harbor porpoises [11 female:10 male]), of which 85 (89%) animals (67 harbor seals/18 harbor porpoises; 24 adults/17 juveniles or subadult/44 pups or calves) successfully yielded 151 bacterial isolates that demonstrated identifiable bacterial growth, representing 26 genera and at least 31 confirmed individual species (Table S1). Eighty-six animals underwent full necropsies, and the remaining were only examined externally at time of sampling. Of the 95 sampled animals, 14 (10 seals, 4 porpoises) were frozen before any sampling took place. Out of these 14 that were frozen before sampling, only one did not yield any bacterial growth from the sampled tissue, in this case, the large intestine. In another animal, an adult harbor seal, the large intestine was sampled pre-freezing, while its other organs were sampled post-thaw at necropsy. For the remaining 13 that were necropsied post-freezing, sampling of organ lesions resulted in bacterial growth. From the 151 isolates, antibiotic resistance determination was performed on 144 (95%) (111 from seals/33 from porpoises). Seven isolates were not tested due to lab limitations (e.g., inability to regrow isolate) or testing was not applicable. Of the 144 isolates tested for resistance, 37% were resistant to at least one antibiotic, 26% were multi-drug resistant, while 61% were sensitive or intermediate to all antibiotics tested. Antibiotic resistant strains were isolated from both species of marine mammals, with 35% of the 74 seals and 52% of the 21 porpoises tested found to have a bacterial isolate resistant to at least one antibiotic (Figure 2). Multi-drug resistance was observed in 24% and 39% of tested seals and porpoises, respectively.

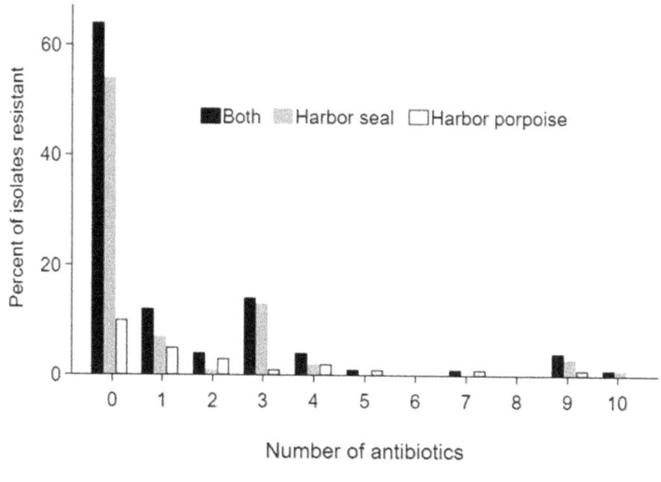

Figure 2. Percentage of bacterial isolates from dead stranded harbor seals and porpoises with antibiotic resistance based on the number of antibiotics to which the isolate was resistant.

Due to the predominance of samples originating from the large intestine, the most frequently cultured bacterium from both marine mammal species was *E. coli* (54%) (Table 1). The next most frequently cultured organisms were beta-hemolytic *Streptococcus* spp. (6%), *Pseudomonas aeruginosa* (3%), *Aeromonas hydrophila* (3%), *Edwardsiella tarda* (3%), *Shewanella algae* (3%), and *Salmonella* spp. (2%). The remaining genera were represented by single isolates. The other most commonly sampled anatomical sites were lung parenchymal lesions noted on gross examination at necropsy, followed by peritoneal fluid in harbor seals presenting with peritonitis (Table 2). There was a significant difference between seals and porpoises in the proportion of isolates that displayed resistance to at least one antibiotic (p-value = 0.004), as well as amongst age classes for both species combined (specifically juveniles compared to pups/calves, p-value = 0.025), but not between the two sampling regions (p-value = 0.248). A significant difference remained between the porpoises and seals when accounting for sampling region (p-value = 0.006). Additional analyses targeting only *E. coli* isolates (n = 112) and only intestinal isolates (n = 100) resulted in similar significant differences. For isolates other than *E. coli* (n = 62), significant differences were noted amongst age classes (p-value = 0.001) but not between species and regions, in proportion of isolates with any resistance.

Table 1. Bacterial species identified from various tissues collected from fresh, dead-stranded harbor seals (*Phoca vitulina*) and harbor porpoises (*Phocoena phocoena*) from the Salish Sea, Washington State, USA. Where applicable, the first number in parentheses refers to the number of isolates resistant to at least one antibiotic, followed by number of isolates resistant to more than one antibiotic. N/A = not tested for resistance.

Organism	Number of Isolates	
	Phoca vitulina	***Phocoena phocoena***
Acinetobacter spp.	0	1 (1/0)
Actinomyces	1 (N/A)	0
Aeromonas hydrophila	2 (2/1)	3 (3/2)
Arcanobacterium haemolyticum	1 (N/A)	0
Arcanobacterium phocae	2 (2/0)	0
β-hemolytic *Streptococcus*	8 (8/8)	1 (1/1)
Buttiauxella agrestis	0	2 (0/0)
Campylobacter spp.	2 (N/A)	0
Escherichia coli	72 (8/4)	10 (1/1)
Edwardsiella hoshinae	1 (0/0)	0
Edwardsiella tarda	4 (0/0)	0
Elizabethkingia meningoseptica	1 (1/1)	0
Enterobacter spp.	1 (1/1)	0
Escherichia fergusonii	2 (2/2)	0
Gamma (γ) hemolytic *Streptococcus*	1 (0/0)	0
Gardnerella vaginalis	1 (1/0)	0
Granulicatella adiacens	1 (N/A)	0
Hafnia alvei	1 (1/1)	0
Moellerella wisconsensis	0	1 (N/A)
Pantoea agglomerans	1 (1/1)	0
Pasteurella group	1 (0/0)	0
Pasteurella multocida	1 (N/A)	0
Photobacterium damselae	0	1 (0/0)
Plesiomonas shigelloides	1 (1/0)	1 (1/0)
Proteus penneri	0	1 (1/1)
Proteus vulgaris	1 (1/1)	0
Providencia rettgeri	0	1 (1/1)
Pseudomonas aeruginosa	5 (5/5)	0
Pseudomonas fluorescens	0	2 (2/2)
Salmonella spp.	3 (3/3)	0
Serratia fonticola	1 (0/0)	0
Shewanella algae	2 (1/0)	2 (2/2)
Shewanella putrefaciens	0	1 (1/1)
Vibrio alginolyticus	0	2 (2/1)
Vibrio cholera	0	1 (0/0)
Vibrio fluvialis	0	1 (1/0)
Vibrio parahaemolyticus	0	2 (2/0)
Vibrio vulnificus	1 (1/1)	0
Total	117	33

Table 2. Bacterial isolates by source and marine mammal species. Harbor seal = *Phoca vitulina;* harbor porpoise = *Phocoena phocoena.* In the case where more than one bacterial species was identified, the number of isolates was greater than the number of swabs taken. Where applicable, the first number in parentheses refers to the number of isolates resistant to at least one antibiotic, followed by number of isolates resistant to more than one antibiotic. N/A = not tested for resistance.

Source	Number of Isolates	
	Phoca vitulina	*Phocoena phocoena*
Abscess (brain)	1 (1/1)	0
Abscess (hind flipper)	2 (1/1)	0
Bronchiole	1 (N/A)	0
Colostrum	2 (1/1)	2 (2/2)
Kidney	1 (0/0)	2 (1/0)
Large intestine	81 (18/12)	22 (12/8)
Lung parenchyma	13 (7/7)	0
Lymph node (unspecified)	2 (2/2)	0
Lymph node (mediastinal)	0	2 (2/1)
Nares	1 (0/0)	0
Oral ulcers	1 (1/0)	0
Peritoneal fluid	9 (3/2)	0
Pleura	0	2 (1/1)
Scapular joint	1 (1/0)	0
Thoracic cavity	0	2 (1/1)
Tonsil	0	2 (1/0)
Uterus	2 (1/1)	0
Total	117	34

The most common species of organisms and patterns of antibiotic resistance obtained from seals and porpoises were markedly dissimilar (Table 3). Combining seals and porpoises together, gram-negative bacteria accounted for a majority (132/151, 87%) of the isolates that were identified, given the intestines were the most sampled source and were most susceptible to enrofloxacin (130/131 isolates, 99%) and marbofloxacin (129/131 isolates, 98%) and least susceptible to amoxicillin (30/130 isolates, 23%) and cephalexin (26/130 isolates, 20%). Only one *E. coli* isolate exhibiting resistance to any antibiotic was recovered from porpoises though all had swabs from the large intestine submitted. Out of the large intestinal swabs submitted from harbor seals that grew *E. coli* colonies (n = 70), only nine (13%) produced isolates that were resistant to at least one antibiotic. Of all the bacteria tested for their antimicrobial susceptibility pattern, the *Pseudomonas* spp. were the most resistant (range: 9–10/15 antibiotics). *Pseudomonas aeruginosa* was isolated from five harbor seals (lung = 3; large intestine = 1; uterus = 1) and *P. fluorescens* from two harbor porpoises (large intestine = 2) (Table 1). They were most susceptible to the aminoglycosides and fluoroquinolones tested, and all but one isolate were also susceptible to doxycycline. One isolate each of *Pseudomonas fluorescens*, *Shewanella algae*, and *Proteus vulgaris* were resistant to imipenem, a member of the carbapenem class of antibiotics.

Table 3. Antibiotic resistance patterns of the most commonly (≥2 isolates) identified bacteria in dead stranded harbor seals (*Phoca vitulina*) and porpoises (*Phocoena phocoena*) from the Salish Sea, Washington, USA. Number at the bottom of each column is the percent resistant out of the total. Gram stain = negative (N) or positive (P).

Bacteria	Gram Stain	Number Isolates	AM	AC	AX	CP	CF	CV	CR	CH	DX	EN	FL	GE	IM	MA	TMS
Phoca vitulina																	
Escherichia coli	N	72			4					1	5	1	1	1		1	3
Beta-hemolytic *Streptococcus*	P	8											8			8	
Pseudomonas aeruginosa	N	5			5	5	5	5	5	5	1		5				5
Salmonella spp.	N	3	3			3								3			
Aeromonas hydrophila	N	2			2	2			1								
Shewanella algae	N	2				1											
Phocoena phocoena																	
Aeromonas hydrophila	N	3			3	2					1						
Pseudomonas fluorescens	N	2		2	2	2	2	2	2	2			2		1		1
Shewanella algae	N	2			1	2					2				1		
Vibrio alginolyticus	N	2		1	1				1								
Vibrio para-haemolyticus	N	2			2												
Total		103	3	3	20	17	7	7	9	8	9	1	16	4	2	9	9

AM = Amikacin, AC = Amoxicillin/clavulanic acid, AX = Amoxicillin, CP = Cephalexin, CF = Cefpodoxime, CV = Cefovecin, CR = Ceftiofur, CH = Chloramphenicol, DX = Doxycycline, EN = Enrofloxacin, GE = Gentamicin, FL = Florfenicol, IM = Imipemem, MA = Marbofloxacin, TMS = Trimethoprim/sulfa.

Beta-hemolytic *Streptococcus* spp. made up more than half (9/13, 56%) of the gram-positive isolates tested for their antibiotic susceptibility. Other gram-positive organisms isolated and tested were *Arcanobacterium phocae* (15%), *Acinetobacter* spp. (8%), and gamma-hemolytic *Streptococcus* (8%). The beta-hemolytic *Streptococcus* spp. isolates were primarily cultured from the lungs (4/9) and were susceptible to most of the antibiotics tested; however, they were all resistant to gentamicin, amikacin, and trimethoprim/sulfamethoxazole. The *Arcanobacterium* isolates were resistant to trimethoprim/sulfamethoxazole and susceptible to the remaining antibiotics.

There was a significant difference in resistance to at least one antibiotic between marine mammal species (p-value = 0.004). Harbor porpoises were at significantly greater (more than three times) risk of having an organism resistant to any antibiotic compared to seals (OR = 3.25; 95% CI: 1.44–7.16). Likewise, a similarly significant difference between the two species was noted when simultaneously accounting for region (p-value = 0.006), but not by region alone, as noted earlier. Similar results were noted when intestinal isolates alone were evaluated. Examining only *E. coli* isolates (n = 82), juvenile harbor porpoises were at greatest risk of resistance compared to calves or adults (OR = 3.34; CI = 1.16–9.65). Marine mammal species and region did not influence *E. coli* isolate resistance significantly. However, in porpoises, *E. coli* isolates were more resistant to the penicillin and cephalosporin antibiotic classes than in seals. Significant differences between age classes, for seals and porpoises combined, were noted for the antibiotics amoxicillin (p-value = 0.023), cephalexin (p-value = 0.019), and cefovecin (p-value = 0.043), specifically for juveniles compared to pups/calves. Pups/calves and juveniles were the source for a majority of the isolates resistant to amoxicillin (19/30, 63%) and cephalexin (15/26, 58%), heavily represented by *Pseudomonas* spp., *Vibrio* spp., and *E. coli* for both antibiotics. No significant differences between the northern and southern study regions by individual antibiotic were observed.

Significant differences in number of isolates (i.e., counts) resistant to antibiotics were not observed between seals and porpoises, study region or age class, individually, nor when all three variables were included in the negative binomial model.

The percentage of total isolates demonstrating resistance to each of the 15 tested antibiotics is shown in Figure 3. Resistance of bacterial isolates to individual antibiotics was significantly greater for porpoises compared to seals for amoxicillin/clavulanic acid (p-value = 0.039; OR = 3.43, 95% CI: 1.06–11.07), amoxicillin (p-value = 0.003; OR = 3.78, CI: 1.58–9.07), cephalexin (p-value = 0.027; OR = 2.83, CI: 1.13–7.10), and cefovecin (p-value = 0.041; OR = 3.93, CI: 1.06–14.54). When analyzing only intestinal isolates, resistance patterns were similar, with the exception there was also significantly greater resistance to cefpodoxime in porpoises (p-value = 0.028). For non-*E. coli* isolates only, significant differences in resistance between species were noted for gentamicin (p-value = 0.039; OR = 9.33, CI: 1.12–77.62) and for trimethoprim/sulfamethoxazole (p-value = 0.011; OR = 7.73, CI: 1.58–37.72), and when adjusted for age class, with no documented resistance to either of the two fluoroquinolones for any of these isolates. When adjusting for study region (north versus south), significant differences between seals and porpoises were again observed for the same four antibiotics. Greater than half (54%) of the 14 isolates that were resistant to amoxicillin/clavulanic acid was attributed to *Pseudomonas* spp., with five in seals and two in porpoises; approximately half of the isolates resistant to amoxicillin and cephalexin were sampled from the large intestine of harbor seals and consisted of a variety of organisms. Four isolates were resistant to imipenem, three of which originated from porpoises.

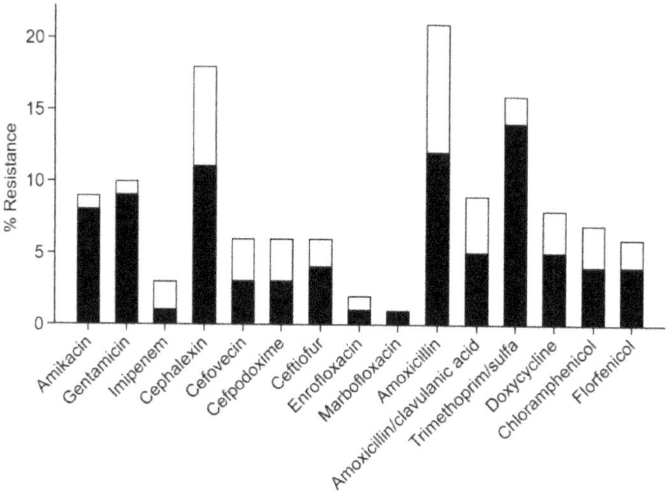

Figure 3. Percentage of bacterial isolates from harbor seals (black) and harbor porpoises (white) displaying resistance to individual antibiotics, grouped by class.

Of the 144 isolates tested for antibiotic sensitivity, 40 (28%) exhibited resistance to multiple antibiotics, with 10% having a multiple antibiotic resistance (MAR) index value > 0.2. An additional calculation of the MAR index value, excluding *Pseudomonas* spp. isolates (n = 7), demonstrated MAR in 23% of 137 isolates, with 20% having an MAR index > 0.2. Of the species-specific isolates tested for antibiotic resistance, 24% (27/111) and 39% (13/33) of seal and porpoise isolates, respectively, were multi-drug resistant (21%; 22/106 and 35%; 11/31, respectively, when *Pseudomonas* excluded). The MAR indices ranged from 0.06 in harbor seal *A. phocae* and *Gardnerella vaginalis* isolates to 0.67 in a *P. aeruginosa* isolate from a seal. The percentage of bacterial isolates within each MAR classification (0, ≤0.2, >2.0) are illustrated by tissue for each marine mammal species (Figure 4). Only tissues represented by at least two isolates were included in the analysis of the MAR index, thus including large intestine, lung parenchyma, peritoneal fluid, colostrum, kidney, lymph node, pleura, thorax, tonsil, uterus, and abscess samples. Isolates with MAR indices > 0.2, as well as no multi-drug resistance, were represented by tissues within the respiratory, reproductive, and lower gastrointestinal systems (lung and pleura, colostrum and uterus, large intestine).

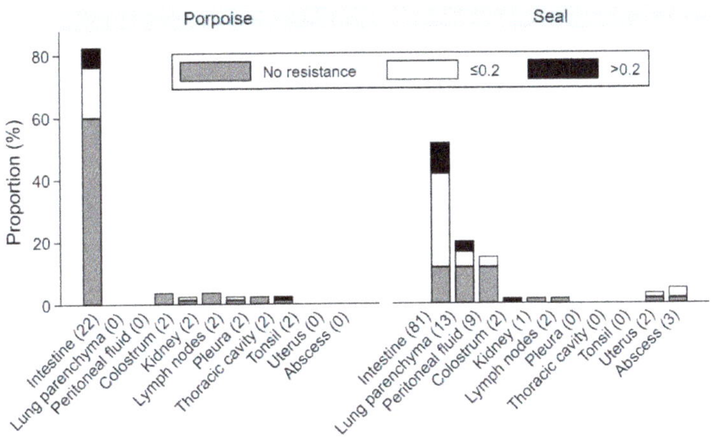

Figure 4. Proportion of bacterial isolates from harbor seals and porpoises with multipleantibiotic resistance originating from individual tissue sources. Proportions are categorized based on their Multiple Antibiotic Resistance Index (MAR): MAR = 0 (no resistance), MAR = $0 \leq 0.2$, or MAR > 0.2. Numbers in parentheses represent number of bacterial isolates from each tissue source.

Though not significant at the alpha level of 0.05, two spatial clusters of multi-drug resistant isolates were observed. One cluster (p-value = 0.081) was located in the central portion of the Salish Sea on Whidbey Island consisting of three harbor seals, two with *P. aeruginosa* and one with *Serratia fonticola*. The other cluster consisted of two harbor porpoises found stranded near the far southern portion of the Salish Sea, from which *Shewanella* spp. were isolated.

4. Discussion

The results of this study demonstrated differences in antibiotic resistance between harbor seals and porpoises inhabiting an urban marine ecosystem in the inland waters of Washington State, near coastlines associated with anthropogenic impacts. Antibiotic resistance was demonstrated in animals stranded throughout the Salish Sea, suggestive of a baseline level of resistance throughout the region. Prior to this study, bacteriologic cultures and antimicrobial susceptibilities were performed on cases submitted as part of necropsy-related sampling, but not consistently throughout the Salish Sea due to variations in stranding response capabilities, logistics, or funding. The results reported here indicate a relatively high level (37%) of antimicrobial resistance in bacteria isolated from stranded marine mammals.

Significant differences were again seen between the two species when the geographic sampling area (north vs. south) was taken into account. Additionally, there were differences in the patterns of antibiotic resistance, or the antibiotic classes to which bacteria demonstrated resistance, between seals and porpoises. Both of these marine mammal species live and forage in nearshore coastal environments, resulting in exposure to sites highly impacted by humans (agricultural runoff, sewage treatment effluent, aquaculture) and the land-to-sea transfer of pathogens [14,37–39]. However, harbor porpoises may differ sufficiently from harbor seals in their habitat use resulting in a greater exposure to anthropogenic pollution. Specifically, porpoises are obligate water dwellers whereas harbor seals spend a portion of their life hauled out of the water to rest and nurse their young. Further work involving molecular techniques will be needed to work out specific mechanisms of resistance between the species but is beyond the scope of this study.

More widespread antibiotic resistance has been reported in stranded marine mammals compared to free-ranging populations, likely due to a bias toward more diseased animals representing in stranded cases [40]. The extent of resistance observed here is consistent with other studies of marine mammals in coastal oceans, noting a relatively high prevalence and similar patterns of antibiotic resistance as presently observed [8,11,13,40]. For example, Lockwood et al. [17] observed that only one antibiotic was observed capable of killing or inhibiting growth of all the isolates tested from harbor seals in the northern Salish Sea, while in bacterial isolates from vertebrates off the northeastern USA coast, 58% of isolates were resistant to at least one antibiotic and 43% to more than one [40]. Isolates (n = 733) originating from bottlenose dolphins (*Tursiops truncatus*) on the east coast of Florida, USA over the periods 2003–2007 and 2010–2015 had an overall 88.2% prevalence of resistance to at least one antibiotic [41].

Organisms such as *Acinetobacter* spp., *A. hydrophila*, *Plesiomonas shigelloides*, *P. fluorescens*, and *Serratia fonticola* are opportunistic in nature and may cause secondary infections in humans and immunocompromised marine mammals such as SRKWs [19]. *E. coli* was the most common isolate from the intestinal tract (n = 69), followed by *Vibrio* spp. (n = 7), consistent with the findings of Stewart et al. (2014) where *Vibrio* spp. (15%) and *E. coli* (6%) were the two most commonly isolated fecal organisms in dolphins. In previous Salish Sea harbor seal studies, *E. coli* was the most consistently isolated organism from sources other than the intestine such as wounds, umbilici, ears, and nares of animals admitted to rehabilitation [17]. Additionally, it was the second most frequently isolated microbe from brain, liver, respiratory tracts, and kidney of free-ranging pups [18], behind *Proteus* spp., which represented only 2/150 (1.3%) isolates in the present study. Enterobacteriaceae are not typically considered primary pathogens but may be secondary opportunistic invaders of preexisting wounds [42]. They were the most frequently isolated gram-negative organisms (56%) in rehabilitating harbor seals in the Salish Sea [17] and were found in 60% of harbor seals live-stranded in California admitted to rehabilitation [8]. While frequently isolated in free-ranging bottlenose dolphins along the east coast of Florida [11,13], Enterobacteriaceae were not the most commonly identified gram-negative bacteria of the present study. They represented only 38% of gram-negative isolates. Ongoing studies are planned to investigate the pathogenic impact of different types of *E. coli* through molecular studies of isolates recovered in this study and to help discover the role of this genus in causing morbidity and mortality in harbor seals, porpoises, and SRKWs. All three *Salmonella* isolates were notably resistant to the aminoglycosides, amikacin and gentamicin, as well as the cephalosporin, cephalexin. This latter drug is of note due to increasing prevalence of multi-drug resistance in *Salmonella*, with special concern to related cephalosporins such as ceftiofur and ceftriaxone [43]. Carbapenem class resistance, seen in isolates of *P. fluorescens*, *S. algae*, and *P. vulgaris*, was accompanied by multi-drug resistance in each of the isolates. Though not treated by carbapenems, marine mammals represent potential reservoirs of multi-drug resistant bacterial strains potentially able to infect humans or other animals [44]. Resistance to carbapenems is an ongoing global public-health problem. This type of antimicrobial resistance, especially when mediated by gene transfer, is spreading rapidly causing serious outbreaks and dramatically limiting treatment options in humans and domestic animals [45,46].

In this study, beta-hemolytic *Streptococcus* was the most common gram-positive isolate and was most frequently recovered from lung tissue (4/9, 44% of beta-hemolytic *Streptococcus* isolates). In rehabilitated Salish Sea harbor seals, it was also the most common gram-positive isolate, frequently found in wounds [17]. Additionally, this organism was isolated from 16% of wounds in live-stranded California harbor seals [8], 21% (4/19 brain and liver isolates) of dead stranded Salish Sea harbor seals [18], and 13 stranded and 3 net caught harbor porpoises (primarily liver and kidney) from the Baltic and North Seas [47].

This contrasts with a study of resistance in seals from the northwest United States Atlantic coast in which *Enterococcus* spp. were the most common gram-positive isolate with 25% (8/32) of isolates originating from the integumentary system [48].

Among groups of bacterial isolates commonly sampled, occurrence of any antibiotic resistance ranged from 10% of isolates (*E. coli*) to 100% (*Pseudomonas* spp., beta-hemolytic *Streptococcus* spp., *Shewanella algae*, *Aeromonas hydrophila*, and *Vibrio parahaemolyticus*) (Table 1), comprising a range of resistance across varying taxonomic groups. It was surprising to observe such low antibiotic resistance in *E. coli* compared to other taxonomic groups, but this was also noted in a large study of antibiotic resistance in seabirds, marine mammals, and fish along the northeast USA [40]. Furthermore, *E. coli* isolated from environmental aquatic samples, that included treated sewage water, rivers, and drinking water, also demonstrated antibiotic resistance [49–51]. Other studies examining resistance in *E. coli* reported higher incidences, ranging from 46% in Tanzanian drinking water [52] to 100% in India [53].

Antibiotic resistance has been reported in aquatic ecosystems contaminated with heavy metals, especially zinc and cadmium, which are thought to contribute to selection of antibiotic-resistant strains such as *E. coli* and *P. aeruginosa* [54]. Heavy metals in the environment may serve as co-selecting agents for antibiotic resistance in human pathogens in reservoirs such as aquatic ecosystems. Heavy metals are not yet quantified in Salish Sea harbor porpoises, but recent studies in regional harbor seals detected significant differences in trace element concentrations among age classes, regions within the Salish Sea, and years sampled [55], and elevated cadmium, copper, and zinc in non-pup seals from the northern vs. southern Salish Sea [56]. Efforts are ongoing to characterize metals in Salish Sea porpoises to inform future antibiotic resistance studies in porpoises.

The greatest proportion of resistant isolates were recovered from juvenile animals, though they were the least represented in number. The absolute counts of resistant isolates amongst the age classes were not significant and may reflect differences in bacterial species within each age group. Most antibiotic resistance studies do not examine differences in susceptibility patterns amongst age classes. Some studies have noted increased antibiotic resistance by methicillin-resistant *Staphylococcus aureus* in older human populations [57]), as well as differences in bacterial species and susceptibility patterns between foals and adult horses [58]. Collecting appropriate proportions of varying age classes for valid comparisons in wildlife studies presents challenges when relying on passive collection of stranded animals. Although not well studied, differences in age class-related resistance between harbor seals and porpoises may reflect differences in microbiome composition such as the nascent fecal biome of pups and calves compared to older animals as has been described in spotted hyenas (*Crocutta crocutta*) [59].

Harbor porpoises, which are more phylogenetically related to the endangered SRKWs, presented with more widespread antibiotic resistance than harbor seals and may represent sentinels for SRKW health. There is overlap in resistant bacteria between harbor porpoises and SRKWs with resistant isolates of *P. fluorescens* and *V. alginolyticus* (from feces in porpoise and breath in SRKW) [19]. Although only eight *E. coli* isolates were obtained from harbor porpoise, they were all sensitive to all antibiotics. They may not have been the same extraintestinal strains documented in SRKWs that were the same clonal lineages (ST73 and ST127) that are often associated with human community-acquired urinary tract disease [20]. Planned genotyping of the present study's isolates will further define their clonal lineages and relationship to SRKW and human isolates.

Multi-drug resistance is an increasingly common problem in human and veterinary medicine and requires a One Health approach to address the multiple facets of health that interplay in forming and promoting resistance [60,61]. For this to happen, we need a deeper, or more complete "One Health" understanding of antimicrobial resistance in free-ranging wildlife, especially regarding the land-sea transfer to marine mammals. Indices of multiple antibiotic resistance (MAR) of $\geq$0.2 were observed in 25/54 (46%) of the resistant bacterial isolates, suggesting seal or porpoise exposure to bacteria from significantly polluted sites.

The influence of anthropogenic activities in aquatic environments, and on their nearshore sediments and soils, may amplify the generation of new antibiotic resistance genes and the spread of resistant bacteria and have serious ramifications on environmental and public health [62,63]. This highlights the importance of the environment as a reservoir of resistance genes and dispersal vectors. The resistance patterns observed in this study suggest that some components of resistance are likely related to environmental origins and may spread without the selective pressure of antibiotic use.

It is unclear how the high levels of antibiotic resistance observed in some of the animals sampled relates to the larger coastal environment since environmental samples were not evaluated or whether the resistance is acquired from other aquatic species. The lack of spatial trends in resistance may reflect the level of sampling, spatial use by the animals sampled, or the complex state of ecosystem connectivity and estuarine exchange flow within the Salish Sea and its coastal shoreline. Connectivity is especially vital in coastal systems where energy and biota are constantly moving and exchanging among ecosystem components [64,65]. Systematic environmental sampling for antibiotic resistance in bacteria throughout the Salish Sea would provide greater insight into determining if antibiotic resistance patterns in marine wildlife are representative of their habitat. Additionally, it might show if human alteration of habitat is fragmenting the region's connectivity or changing normal ecosystem processes that help limit transfer of antibiotic resistant bacteria. For example, Lamb et al. [66] demonstrated that healthy seagrass meadows can reduce the land-sea transfer of harmful bacteria. Seagrass meadows can naturally remove or reduce pathogens in terrestrially-sourced wastewater by up to 50% and reduce coral reef disease, yet in Washington's half of the Salish Sea, localized small-scale eelgrass (*Zostera marina*) declines have been significant [67].

The findings from this study are significant for marine mammal and human health. Since marine mammals are apex predators, with their health reflective of their shared marine ecosystem with humans, they serve as sentinels of ecosystem health [68]. This is particularly key in a region such as the Salish Sea where much of the human population lives along the coastline. The patterns of antibiotic resistance described in two Salish Sea marine mammal populations may indicate a potential public health risk since disease related to these bacteria may coincide with emerging infectious diseases in these mammals resulting from environmental perturbations or increasing resistance in aquatic species in general. Study results will help inform local human and veterinary health officials, as well as raise public awareness of drug resistance in the marine environment.

Though zoonotic pathogens from marine mammals are not widely distributed in human populations, aquatic mammals are known to carry pathogens that pose a risk to human health [69,70]. Importantly, several bacteria were identified in this study that can cause infections of public health significance and are considered reportable to the Centers for Disease Control and Prevention (CDC) [71], including *Vibrio cholerae*, *V. parahaemolyticus*, and *V. vulnificus*. *Vibrio* species are abundant and ubiquitous in the marine environment and are a frequent cause of gastrointestinal illness in humans associated with seafood consumption (*V. parahaemolyticus*) in the Pacific Northwest [72] and wound infections resulting in high morbidity and mortality (*V. vulnificus*) [73]. An isolate of *V. cholerae* was identified from the large intestine of a harbor porpoise calf found in the northern portion of the study area near the San Juan Islands, but characterized as non-toxigenic O1, not O139, the cause of reportable cholera [71]. A survey of estuarine waters of Washington, Oregon, and California detected non-O1 strains of V. cholerae in Washington [74]. That study posited that the low incidence of non-O1 *Vibrio* strains detected suggested a potential for human, and thus marine mammal, infection; however, the low frequency of toxigenic strains indicates a lower threat to animals in the Pacific Northwest compared to other regions such as the Gulf of Mexico and Atlantic coast. Other potential sources of exposure to zoonotic bacteria from marine mammals include occupational exposure [69,75], touching live or dead animals on the beach, and consumption of marine mammals [76].

A confounding factor of this study is that most of the samples were fecal swabs obtained via rectal sampling or at necropsy due to ease of procurement, and as previously described, most of the bacteria isolated from the fecal swabs were *E. coli*, which showed relatively low incidence of antibiotic resistance compared to other tissues and other bacterial groups. Thus, it is possible that the differences between the seals and porpoises and age groups may have been due to the sample types collected; however, these were controlled for in the analyses and thus do not exert an undue influence on the analytical results. Sampling of previously frozen carcasses (14 in total) might have hindered the ability to isolate bacteria (and thus resulting resistance patterns). Effort was made to minimize sampling post-thaw, but was not always avoidable. The greatest impact on the analyses was a reduction in the number and types of bacterial isolates that would have been obtained if the post-thaw animals had been sampled prior to freezing. The methods used in this study were limited to aerobic culture of organisms using commercial instruments and databases, and a limited number of antibiotics, and was not intended to investigate the complete microbiomes of either harbor seals or porpoises. Rather, it was designed to evaluate samples of convenience for bacterial prevalence and sensitivities that are of SRKW and public health relevance, although tested samples likely contained additional bacterial species that were present but not speciated due to lack of growth from suboptimal storage or other extrinsic factors, resulting in underreporting of prevalence. High-throughput genome sequencing methods, and more recently DNA-based metagenomics, have revolutionized the ability to quickly look at dominant bacteria and antibiotic resistance genes (resistomes) to determine the relative contribution and importance of a bacterial species to an animal's microbial community [19,20,77,78]. Though resistomes were not examined in this study, likely resulting in antibiotic resistance genes being missed, future studies should include genome sequencing in addition to traditional culture methods.

5. Conclusions

In this study, we document widespread antibiotic resistance in bacterial isolates from two species of marine mammals, with more marked multi-drug resistance in harbor porpoises compared to harbor seals. The high incidence of single and multi-drug resistance was consistent with other studies in marine species. Results of this study also confirm previous work suggesting multi-drug resistance may be common in bacteria originating in marine mammals. The novelty of this study is its focus on two species with overlapping ranges but dissimilar ecological niches (completely aquatic versus semi-aquatic). We provided an initial simultaneous glimpse at resistance patterns in multiple regions and age classes within an urbanized marine ecosystem that may serve as a reservoir for antibiotic resistance. These two marine mammal species live in nearshore waters of the Salish Sea and likely come into regular contact with humans and their associated activities. Thus, the relatively high occurrence of antibiotic resistance may reflect a large environmental reservoir of antibiotic resistant organisms occurring in this body of water, or inherent differences in resistance patterns or susceptibility to resistance genes between the two species. Due to the large geographic area of the Salish Sea and the variety of its anthropogenic activities, further investigation of temporal and spatial resistance patterns, as well as environmental sampling, will better inform natural resource managers working to recover endangered SRKWs as well as public health officials in the region. By engaging the human, animal, and environmental health sectors together to monitor antibiotic resistance patterns in marine species, efforts to address antibiotic resistance in a collaborative One Health approach will benefit this urban ecosystem and its inhabitants.

Supplementary Materials: The following is available online at http://www.mdpi.com/xxx/s1. Table S1: Raw data results from sampling of harbor seals and harbor porpoises for aerobic bacterial culture and sensitivity.

Author Contributions: Conceptualization, S.A.N.; methodology, S.A.N., D.M.L., and J.L.H.; software, S.A.N.; validation, S.A.N., D.M.L., J.L.H., and J.K.G.; formal analysis, S.A.N.; investigation, S.A.N.,

J.K.G., J.L.H., D.M.L., J.K.O., V.S., A.E., A.S., S.B., S.D., C.K., B.C., M.J., and R.M.; resources, S.A.N., J.K.G., J.L.H., D.M.L., J.K.O., V.S., A.E., A.S., S.B., S.D., C.K., B.C., M.J., and R.M.; data curation, S.A.N.; writing—original draft preparation, S.A.N.; writing—review and editing, S.A.N., J.K.G., J.L.H., D.M.L., J.K.O., V.S., A.E., A.S., S.B., S.D., C.K., B.C., M.J., and R.M.; visualization, S.A.N. and J.K.G.; supervision, S.A.N.; project administration, S.A.N. and S.B.; funding acquisition, S.A.N., S.B., S.D., J.L.H., V.S., C.K., J.K.O., and B.C. All authors have read and agreed to the published version of the manuscript.

Funding: Funds to support this project were provided by the generous donations of more than 148 public citizens submitted through Experiment (www.experiment.com), a web-based source for crowdfunding science-based research projects, and from the Wildlife Disease Association. Efforts to support collection of stranded marine mammals were partially funded through the John H. Prescott Marine Mammal Rescue Assistance Grant Program.

Institutional Review Board Statement: IACUC approval was not required by any institutions or organizations for this work, as all carcasses used in this study were found dead prior to inclusion in the study. Authorization to collect dead, beach-cast marine mammals is given to each collaborative organization in this study through a permit from the United States National Marine Fisheries Service Marine Mammal Health and Stranding Response Program (#18786-04).

Data Availability Statement: The data generated in this study are available in Table S1.

Acknowledgments: Study animals were collected with the assistance of volunteers and staff from Orca Network/Central Puget Sound Marine Mammal Stranding Network (specifically Garry Heinrich, Bob and Kelly Post, Mary Jo Adams, Jill Hein, and Matt Klope), Washington Department of Fish and Wildlife, Cascadia Research Collective (specifically Erin Keene), San Juan County Marine Mammal Stranding Network (The Whale Museum), Seal Sitters Marine Mammal Stranding Network (especially Lynn Shimamoto), Whatcom County Marine Mammal Stranding Network, Port Townsend Marine Science Center/East Jefferson County Marine Mammal Stranding Network, Sno-King Marine Mammal Rescue, MaST Center Marine Mammal Stranding Team at Highline College, and World Vets. Meridith Grogan, Meghan Riddell, and Ronda Reisdorph (Phoenix Laboratory, Mukilteo, Washington), and Linda Rhodes (NOAA Fisheries, Northwest Fisheries Science Center, Seattle, Washington) for help handling isolates, and Cindy Wu (Experiment.com) for technical help with project crowdsourcing. We appreciate the suggestions of four anonymous reviewers that improved the manuscript.

Conflicts of Interest: The authors declare no conflict of interest.

References

1. McFall-Ngai, M.; Hadfield, M.G.; Bosch, T.C.G.; Carey, H.V.; Domazet-Lošo, T.; Douglas, A.E.; Dubilier, N.; Eberl, G.; Fukami, T.; Gilbert, S.F.; et al. Animals in a bacterial world, a new imperative for the life sciences. *Proc. Natl. Acad. Sci. USA* **2013**, *110*, 3229–3236. [CrossRef] [PubMed]
2. World Health Organization. Antibiotic Resistance. Available online: http://www.who.int/news-room/factsheets/detail/antibiotic-resistance (accessed on 27 May 2020).
3. Robinson, T.P.; Bu, D.P.; Carrique-Mas, J.; Fèvre, E.M.; Gilbert, M.; Grace, D.; Hay, S.I.; Jiwakanon, J.; Kakkar, M.; Kariuki, S.; et al. Antibiotic resistance is the quintessential One Health issue. *Trans. R. Soc. Trop. Med. Hyg.* **2016**, *110*, 377–380. [CrossRef]
4. National Academies of Sciences, Engineering, and Medicine. *Combating Antimicrobial Resistance: A One Health Approach to a Global Threat: Proceedings of a Workshop*; The National Academies Press: Washington, DC, USA, 2017.
5. Smith, D.L.; Harris, A.D.; Johnson, J.A.; Silbergeld, E.K.; Morris, J.J.G. Animal antibiotic use has an early but important impact on the emergence of antibiotic resistance in human commensal bacteria. *Proc. Natl. Acad. Sci. USA* **2002**, *99*, 6434–6439. [CrossRef] [PubMed]
6. Stewart, J.R.; Gast, R.J.; Fujioka, R.S.; Solo-Gabriele, H.M.; Meschke, J.S.; Amaral-Zettler, L.; Del Castillo, E.; Polz, M.F.; Collier, T.K.; Strom, M.; et al. The coastal environment and human health: Microbial indicators, pathogens, sentinels and reservoirs. *Environ. Health* **2008**, *7*, S3. [CrossRef] [PubMed]
7. Silbergeld, E.K.; Graham, J.; Price, L.B. Industrial Food Animal Production, Antimicrobial Resistance, and Human Health. *Annu. Rev. Public Health* **2008**, *29*, 151–169. [CrossRef]
8. Johnson, S.P.; Nolan, S.; Gulland, F.M. Antimicrobial susceptibility of bacteria isolated from pinnipeds stranded in central and northern California. *J. Zoo Wildl. Med.* **1998**, *29*, 288–294.
9. Miranda, C.D.; Zemelman, R. Antibiotic resistance in fish from the Concepción Bay, Chile. *Mar. Pollut. Bull.* **2001**, *42*, 1096–1102. [CrossRef]

10. Greig, T.W.; Bemiss, J.A.; Lyon, B.R.; Bossart, G.D.; Fair, P.A. Prevalence and diversity of antibiotic-resistant Escherichia coli in bottlenose dolphins (Tursiops truncatus) from the Indian River Lagoon, Florida and Charleston Harbor area, South Caro-lina. *Aquat. Mamm.* **2007**, *33*, 185–194. [CrossRef]
11. Schaefer, A.M.; Goldstein, J.D.; Reif, J.S.; Fair, P.A.; Bossart, G.D. Antibiotic-resistant organisms cultured from Atlantic bot-tlenose dolphins (*Tursiops truncatus*) inhabiting estuarine waters of Charleston, SC, and Indian River Lagoon, FL. *Ecohealth* **2009**, *6*, 33–41. [CrossRef]
12. Blackburn, J.K.; Mitchell, M.A.; Blackburn, M.-C.H.; Curtis, A.; Thompson, B.A. Evidence of Antibiotic Resistance in Free-Swimming, Top-Level Marine Predatory Fishes. *J. Zoo Wildl. Med.* **2010**, *41*, 7–16. [CrossRef]
13. Stewart, J.R.; Townsend, F.; Lane, S.M.; Dyar, E.; Hohn, A.; Rowles, T.K.; Staggs, L.; Wells, R.S.; Balmer, B.C.; Schwacke, L.H. Survey of antibiotic-resistant bacteria isolated from bottlenose dolphins Tursiops truncatus in the southeastern USA. *Dis. Aquat. Org.* **2014**, *108*, 91–102. [CrossRef] [PubMed]
14. Oates, S.C.; Miller, M.A.; Byrne, B.A.; Chouicha, N.; Hardin, D.; Jessup, D.; Dominik, C.; Roug, A.; Schriewer, A.; Jang, S.S.; et al. Epidemiology and potential land-sea transfer of enteric bacteria from terrestrial to marine species in the monterey bay region of california. *J. Wildl. Dis.* **2012**, *48*, 654–668. [CrossRef] [PubMed]
15. Michael, I.; Rizzo, L.; McArdell, C.S.; Manaia, C.M.; Merlin, C.; Schwartz, T.; Dagot, C.; Fatta-Kassinos, D. Urban waste water treatment plants as hotspots for the release of antibiotics in the environment: A review. *Water Res.* **2013**, *47*, 957–993. [CrossRef] [PubMed]
16. Hatosy, S.M.; Martiny, A.C. The Ocean as a Global Reservoir of Antibiotic Resistance Genes. *Appl. Environ. Microbiol.* **2015**, *81*, 7593–7599. [CrossRef]
17. Lockwood, S.K.; Chovan, J.L.; Gaydos, J.K. Aerobic bacterial isolations from harbor seals (*Phoca vitulina*) stranded in Wash-ington: 1992–2003. *J. Zoo Wildl. Med.* **2006**, *37*, 281–291. [CrossRef]
18. Steiger, G.H.; Calambokidis, J.; Cubbage, J.C.; Skilling, D.E.; Smith, A.W.; Gribble, D.H. Mortality of harbor seal pups at different sites in the inland waters of washington. *J. Wildl. Dis.* **1989**, *25*, 319–328. [CrossRef]
19. Raverty, S.A.; Rhodes, L.D.; Zabek, E.; Eshghi, A.; Cameron, C.E.; Hanson, M.B.; Schroeder, J.P. Respiratory microbiome of endangered southern resident killer whales and microbiota of surrounding sea surface microlayer in the eastern North Pa-cific. *Sci. Rep.* **2017**, *7*, 394. [CrossRef]
20. Melendez, D.; Roberts, M.C.; Greninger, A.L.; Weissman, S.; No, D.; Rabinowitz, P.; Wasser, S. Whole-genome analysis of extraintestinal pathogenic Escherichia coli (ExPEC) MDR ST73 and ST127 isolated from endangered southern resident killer whales (Orcinus orca). *J. Antimicrob. Chemother.* **2019**, *74*, 2176–2180. [CrossRef]
21. Geraci, J.R.; Lounsbury, V.J. *Marine Mammals Ashore: A Field Guide for Strandings*, 2nd ed.; E. John Schmitz & Sons. Inc: Sparks, MD, USA, 2005; pp. 177–178.
22. Norman, S.A.; Bowlby, C.E.; Brancato, M.S.; Calambokidis, J.; Duffield, D.; Gearin, P.J.; Gornall, T.A.; Gosho, M.E.; Hanson, B.; Hodder, J.; et al. Cetacean strandings in Oregon and Washington between 1930 and 2002. *J. Cetacean Res. Manag.* **2004**, *6*, 87–99.
23. Warlick, A.J.; Duffield, D.A.; Lambourn, D.M.; Jeffries, S.J.; Rice, J.M.; Gaydos, J.K.; Huggins, J.L.; Calambokidis, J.; Lahner, L.L.; Olson, J.; et al. Spatio-Temporal Characterization of Pinniped Strandings and Human Interaction Cases in the Pacific Northwest, 1991–2016. *Aquat. Mamm.* **2018**, *44*, 299–318. [CrossRef]
24. Carretta, J.V.; Forney, K.A.; Oleson, E.M.; Weller, D.W.; Lang, A.R.; Baker, J.D.; Muto, M.M.; Hanson, B.; Orr, A.J.; Huber, H.R.; et al. *U.S. Pacific Marine Mammal Stock Assessments: 2018. NOAA Technical Memorandum NMFS-SWFSC-617*; U.S. De-partment of Commerce: Washington, DC, USA, 2019.
25. Cottrell, P.E.; Jeffries, S.; Beck, B.; Ross, P.S. Growth And Development In Free-Ranging Harbor Seal (Phoca Vitulina) Pups From Southern British Columbia, Canada. *Mar. Mammal Sci.* **2002**, *18*, 721–733. [CrossRef]
26. Gearin, P.J.; Melin, S.R.; Delong, R.L.; Kajimura, H.; Johnson, M.A. Harbor porpoise interactions with a chinook salmon set-net fishery in Washington State. *Rep. Int. Whal. Commn. Special* **1994**, *15*, 427–438.
27. Krogh, J.; Ianson, D.; Hamme, R.C.; Lowe, C.J. Risks of hypoxia and acidification in the high energy coastal environment near Victoria, Canada's untreated municipal sewage outfalls. *Mar. Pollut. Bull.* **2018**, *133*, 517–531. [CrossRef]
28. Bonham, P.A. Swab cultures for diagnosing wound infections: A literature review and clinical guideline. *J. Wound Ostomy. Cont.* **2009**, *36*, 389–395. [CrossRef] [PubMed]
29. Quinn, P.J.; Markey, B.K.; Leonard, F.C.; FitzPatrick, E.S.; Fanning, S.; Hartigan, P.J. *Veterinary Microbiology and Microbial Disease*, 2nd ed.; Wiley-Blackwell: West Sussex, UK, 2011.
30. Sweeney, M.T.; Lubbers, B.V.; Schwarz, S.; Watts, J.L. Applying definitions for multidrug resistance, extensive drug re-sistance and pandrug resistance to clinically significant livestock and companion animal bacterial pathogens. *J. Antimicrob. Chemother.* **2018**, *73*, 1460–1463. [CrossRef]
31. Krumperman, P.H. Multiple antibiotic resistance indexing of Escherichia coli to identify high-risk sources of fecal contamina-tion of foods. *Appl. Environ. Microbiol.* **1983**, *46*, 165–170. [CrossRef]
32. Chitanand, M.P.; Kadam, T.A.; Gyananath, G.; Totewad, N.D.; Balhal, D.K. Multiple antibiotic resistance indexing of coliforms to identify high risk contamination sites in aquatic environment. *Indian J. Microbiol.* **2010**, *50*, 216–220. [CrossRef]
33. Titilawo, Y.; Sibanda, T.; Obi, L.; Okoh, A. Multiple antibiotic resistance indexing of Escherichia coli to identify high-risk sources of faecal contamination of water. *Environ. Sci. Pollut. Res.* **2015**, *22*, 10969–10980. [CrossRef]

34. Tambekar, D.H.; Hirulkar, N.B.; Waghmare, A.S. MAR indexing to discriminate the source of faecal contamination in drinking water. *Nat. Environ. Pollut. Tech.* **2005**, *4*, 525–528. [CrossRef]
35. Kaneene, B.J.; Miller, R.; Sayah, R.; Johnson, Y.J.; Gilliland, D.; Gardiner, J.C. Considerations when using discrimination function analysis of antimicrobial resistance profiles to identify sources of faecal contamination of surface water in Michi-gan. *Appl. Environ. Microbiol.* **2007**, *73*, 2878–2890. [CrossRef]
36. Kulldorff, M. A spatial scan statistic. *Commun. Stat. Theory Methods* **1997**, *26*, 1481–1496. [CrossRef]
37. Smaldone, G.; Marrone, R.; Cappiello, S.; Martin, G.A.; Oliva, G.; Cortesi, M.L.; Anastasio, A. Occurrence of antibiotic re-sistance in bacteria isolated from seawater organisms caught in Campania Region: Preliminary study. *BMC Vet. Res.* **2014**, *10*, 161. [CrossRef] [PubMed]
38. Watts, J.; Schreier, H.J.; Lanska, L.; Hale, M.S. The Rising Tide of Antimicrobial Resistance in Aquaculture: Sources, Sinks and Solutions. *Mar. Drugs* **2017**, *15*, 158. [CrossRef] [PubMed]
39. Ramey, A.M.; Ahlstrom, C.A. Antibiotic Resistant Bacteria in Wildlife: Perspectives on Trends, Acquisition and Dissemination, Data Gaps, and Future Directions. *J. Wildl. Dis.* **2020**, *56*, 1–15. [CrossRef] [PubMed]
40. Rose, J.M.; Gast, R.J.; Bogomolni, A.; Ellis, J.C.; Lentell, B.J.; Touhey, K.; Moore, M. Occurrence and patterns of antibiotic re-sistance in vertebrates off the Northeastern United States coast. *FEMS Microbiol. Ecol.* **2009**, *67*, 421–431. [CrossRef]
41. Schaefer, A.M.; Bossart, G.D.; Harrington, T.; Fair, P.A.; McCarthy, P.J.; Reif, J.S. Temporal Changes in Antibiotic Resistance Among Bacteria Isolated from Common Bottlenose Dolphins (*Tursiops truncatus*) in the Indian River Lagoon, Florida, 2003–2015. *Aquat. Mamm.* **2019**, *45*, 533–542. [CrossRef]
42. Higgins, R. Bacteria and fungi of marine mammals: A review. *Can. Veter. J.* **2000**, *41*, 105–116.
43. Hur, J.; Jawale, C.; Lee, J.H. Antimicrobial resistance of Salmonella isolated from food animals: A review. *Food Res. Int.* **2012**, *45*, 819–830. [CrossRef]
44. Haenni, M.; Bour, M.; Châtre, P.; Madec, J.-Y.; Plésiat, P.; Jeannot, K. Resistance of Animal Strains of Pseudomonas aeruginosa to Carbapenems. *Front. Microbiol.* **2017**, *8*, 1847. [CrossRef]
45. Meletis, G. Carbapenem resistance: Overview of the problem and future perspectives. *Ther. Adv. Infect. Dis.* **2015**, *3*, 15–21. [CrossRef]
46. Smith, A.; Wayne, A.; Fellman, C.L.; Rosenbaum, M.H. Usage patterns of carbapenem antimicrobials in dogs and cats at a veterinary tertiary care hospital. *J. Veter. Intern. Med.* **2019**, *33*, 1677–1685. [CrossRef] [PubMed]
47. Swenshon, M.; Lammler, C.; Siebert, U. Identification and Molecular Characterization of Beta-Hemolytic Streptococci Isolated from Harbor Porpoises (Phocoena phocoena) of the North and Baltic Seas. *J. Clin. Microbiol.* **1998**, *36*, 1902–1906. [CrossRef] [PubMed]
48. Wallace, C.C.; Yund, P.O.; Ford, T.E.; Matassa, K.A.; Bass, A.L. Increase in Antimicrobial Resistance in Bacteria Isolated from Stranded Marine Mammals of the Northwest Atlantic. *EcoHealth* **2013**, *10*, 201–210. [CrossRef] [PubMed]
49. Boon, P.I.; Cattanach, M. Antibiotic resistance of native and faecal bacteria isolated from rivers, reservoirs and sewage treatment facilities in Victoria, south-eastern Australia. *Lett. Appl. Microbiol.* **1999**, *28*, 164–168. [CrossRef]
50. Lima-Bittencourt, C.I.; Cursino, L.; Gonçalves-Dornelas, H.; Pontes, D.S.; Nardi, R.M.D.; Callisto, M.; Chartone-Souza, E.; Nascimento, A.M. Multiple antimicrobial resistance in Enterobacteriaceae isolates from pristine freshwater. *Genet. Mol. Res.* **2007**, *6*, 510–521.
51. Coleman, B.L.; Louie, M.; Salvadori, M.I.; McEwen, S.A.; Neumann, N.; Sibley, K.; Irwin, R.J.; Jamieson, F.B.; Daignault, D.; Majury, A.; et al. Contamination of Canadian private drinking water sources with antimicrobial resistant Escherichia coli. *Water Res.* **2013**, *47*, 3026–3036. [CrossRef]
52. Lyimo, B.; Buza, J.; Subbiah, M.; Smith, W.A.; Call, D.R. Comparison of antibiotic resistant Escherichia coli obtained from drinking water sources in northern Tanzania: A cross-sectional study. *BMC Microbiol.* **2016**, *16*, 1–10. [CrossRef]
53. Pantak, S.P.; Gopal, K. Prevalence of bacterial contamination with antibiotic resistant and enterotoxigenic fecal coliforms in treated drinking water. *J. Toxicol. Environ. Health A* **2008**, *71*, 427–433. [CrossRef]
54. Nguyen, C.C.; Hugie, C.N.; Kile, M.L.; Navab-Daneshmand, T. Association between heavy metals and antibiotic-resistant human pathogens in environmental reservoirs: A review. *Front. Environ. Sci. Eng.* **2019**, *13*, 46. [CrossRef]
55. Akmajian, A.; Calambokidis, J.; Huggins, J.L.; Lambourn, D.M. Age, region, and temporal patterns of trace elements meas-ured in stranded harbor seals (Phoca vitulina richardii) from Washington inland waters. *Northwest Nat.* **2014**, *95*, 83–91. [CrossRef]
56. Ashley, E.A.; Olson, J.K.; Raverty, S.; Wilkinson, K.; Gaydos, J.K. Trace Element Concentrations in Livers of Pacific Harbor Seals (Phoca vitulina richardii) from San Juan County, Washington, USA. *J. Wildl. Dis.* **2020**, *56*, 429. [CrossRef] [PubMed]
57. Garcia, A.; Delorme, T.; Nasr, P. Patient age as a factor of antibiotic resistance in methicillin-resistant Staphylococcus aureus. *J. Med. Microbiol.* **2017**, *66*, 1782–1789. [CrossRef] [PubMed]
58. Miagkoff, L.; Archambault, M.; Bonilla, A.G. Antimicrobial susceptibility patterns of bacterial isolates cultured from syno-vial fluid samples from horses with suspected septic synovitis: 108 cases (2008–2017). *J. Am. Vet. Med. Assoc.* **2020**, *256*, 800–807. [CrossRef] [PubMed]
59. Rojas, C.A.; Holekamp, K.E.; Winters, A.D.; Theis, K.R. Body site-specific microbiota reflect sex and age-class among wild spotted hyenas. *FEMS Microbiol. Ecol.* **2020**, *96*, 1–14. [CrossRef] [PubMed]
60. Suzuki, S.; Pruden, A.; Virta, M.; Zhang, T. Editorial: Antibiotic Resistance in Aquatic Systems. *Front. Microbiol.* **2017**, *8*, 14. [CrossRef]

61. Frey, E. The role of companion animal veterinarians in one-health efforts to combat antimicrobial resistance. *J. Am. Veter. Med. Assoc.* **2018**, *253*, 1396–1404. [CrossRef]
62. Rizzo, L.; Manaia, C.; Merlin, C.; Schwartz, T.; Dagot, C.; Ploy, M.C.; Michael, I.; Fatta-Kassinos, D. Urban wastewater treat-ment plants as hotspots for antibiotic resistant bacteria and genes spread into the environment: A review. *Sci. Total Environ.* **2013**, *447*, 345–360. [CrossRef]
63. Ramirez-Castillo, F.Y.; Harel, J.; Moreno-Flores, A.C.; Loera-Muro, A.; Guerrero-Barrera, A.L.; Avelar-González, F.J. Anti-microbial resistance: The role of aquatic environments. *Int. J. Curr. Res. Acad. Rev.* **2014**, *2*, 231–246.
64. Gaydos, J.K.; Dierauf, L.; Kirby, G.; Brosnan, D.; Gilardi, K.; Davis, G.E. Top 10 principles for designing healthy coastal eco-systems like the Salish Sea. *Ecohealth* **2008**, *5*, 460. [CrossRef]
65. Fresh, K.; Dethier, M.; Simenstad, C.; Logsdon, M.; Shipman, H.; Tanner, C.D.; Leschine, T.M.; Mumford, T.F.; Gelfenbaum, G.; Shuman, R.; et al. Implications of observed anthropogenic changes to the nearshore ecosystems in Puget Sound. Puget Sound Nearshore Ecosystem Restoration Project. *Tech. Rep.* **2011**. Available online: http://www.pugetsoundnearshore.org/technical_papers/implications_of_observed_ns_change.pdf (accessed on 24 June 2020).
66. Lamb, J.B.; Van De Water, J.A.J.M.; Bourne, D.G.; Altier, C.; Hein, M.Y.; Fiorenza, E.A.; Abu, N.; Jompa, J.; Harvell, C.D. Seagrass ecosystems reduce exposure to bacterial pathogens of humans, fishes, and invertebrates. *Science* **2017**, *355*, 731–733. [CrossRef]
67. Shelton, A.O.; Francis, T.B.; Feist, B.E.; Williams, G.D.; Lindquist, A.; Levin, P.S. Forty years of seagrass population stability and resilience in an urbanizing estuary. *J. Ecol.* **2016**, *105*, 458–470. [CrossRef]
68. Bossart, G.D. Marine Mammals as Sentinel Species for Oceans and Human Health. *Veter. Pathol.* **2010**, *48*, 676–690. [CrossRef] [PubMed]
69. Moore, M.J.; Gast, R.J.; Bogomolni, A.L. Marine vertebrate zoonoses: An overview of the DAO Special Issue. *Dis. Aquat. Organ.* **2008**, *81*, 1–3. [CrossRef] [PubMed]
70. Waltzek, T.B.; Cortés-Hinojosa, G.; Wellehan, J.F.X., Jr.; Gray, G.C. Marine mammal zoonoses: A review of disease manifes-tations. *Zoonoses Public Health* **2012**, *59*, 521–535. [CrossRef]
71. US Centers for Disease Control and Prevention (CDC), National Notifiable Diseases Surveillance System. Available online: https://wwwn.cdc.gov/nndss/ (accessed on 26 May 2020).
72. Paranjpye, R.N.; Nilsson, W.; Liermann, M.; Hilborn, E.D.; George, B.J.; Li, Q.; Bill, B.D.; Trainer, V.L.; Strom, M.S.; Sandifer, P.A. Environmental influences on the seasonal distribution of Vibrio parahaemolyticus in the Pacific Northwest of the USA. *FEMS Microbiol. Ecol.* **2012**, *91*, fiv121. [CrossRef]
73. Strom, M.S.; Paranjpye, R.N. Epidemiology and pathogenesis of Vibrio vulnificus. *Microbes Infect.* **2000**, *2*, 177–188. [CrossRef]
74. Kaysner, C.A.; Abeyta, C.; Wekell, M.M.; DePaola, A.; Stott, R.F.; Leitch, J.M. Incidence of Vibrio cholerae from estuaries of the United States West Coast. *Appl. Environ. Microbiol.* **1987**, *53*, 1344–1348. [CrossRef]
75. Hunt, T.; Ziccardi, M.; Gulland, F.; Yochem, P.; Hird, D.; Rowles, T.; Mazet, J. Health risks for marine mammal workers. *Dis. Aquat. Org.* **2008**, *81*, 81–92. [CrossRef]
76. Tryland, M. Zoonoses of arctic marine mammals. *Infect. Dis. Rev.* **2000**, *2*, 55–64.
77. Reuter, J.A.; Spacek, D.V.; Snyder, M.P. High-Throughput Sequencing Technologies. *Mol. Cell* **2015**, *58*, 586–597. [CrossRef]
78. Chen, Y.-M.; Holmes, E.C.; Chen, X.; Tian, J.-H.; Lin, X.-D.; Qin, X.-C.; Gao, W.-H.; Liu, J.; Wu, Z.; Zhang, Y.-Z. Diverse and abundant resistome in terrestrial and aquatic vertebrates revealed by transcriptional analysis. *Sci. Rep.* **2020**, *10*, 1–11. [CrossRef]

oceans

MDPI

Article

An Expert Elicitation of the Effects of Low Salinity Water Exposure on Bottlenose Dolphins

Cormac Booth [1,*] and Len Thomas [2]

[1] SMRU Consulting, Scottish Oceans Institute, University of St Andrews, East Sands, Fife KY16 8LB, UK
[2] Centre for Research into Ecological & Environmental Modelling, University of St Andrews, The Observatory, Buchanan Gardens, St Andrews KY16 9LZ, UK; len.thomas@st-andrews.ac.uk
* Correspondence: cgb@smruconsulting.com; Tel.: +44-1314638555

Abstract: There is increasing concern over anthropogenically driven changes in our oceans and seas, from a variety of stressors. Such stressors include the increased risk of storms and precipitation, offshore industries and increased coastal development which can affect the marine environment. For some coastal cetacean species, there is an increased exposure to low salinity waters which have been linked with a range of adverse health effects in bottlenose dolphins. Knowledge gaps persist regarding how different time–salinity exposures affect the health and survival of animals. In such data-poor instances, expert elicitation can be used to convert an expert's qualitative knowledge into subjective probability distributions. The management implications of this stressor and the subjective nature of expert elicitation requires transparency; we have addressed this here, utilizing the Sheffield Elicitation Framework. The results are a series of time response scenarios to estimate time to death in bottlenose dolphins, for use when data are insufficient to estimate probabilistic summaries. This study improves our understanding of how low salinity exposure effects dolphins, guiding priorities for future research, while its outputs can be used to support coastal management on a global scale.

Keywords: freshwater; cetacean; *Tursiops* sp.; wildlife management; marine biology; salinity; human disturbance; dose response

Citation: Booth, C.; Thomas, L. An Expert Elicitation of the Effects of Low Salinity Water Exposure on Bottlenose Dolphins. *Oceans* **2021**, *2*, 179–192. https://doi.org/10.3390/oceans2010011

Academic Editor: Alexander Werth

Received: 3 December 2020
Accepted: 9 February 2021
Published: 14 February 2021

1. Introduction

Over recent decades, there has been increasing concern over anthropogenically driven changes in our oceans and seas representing both single and multiple stressors for marine species [1]. These stressors exist across a range of scales, ranging from the pressing concern of climate change [2], marine heat waves [3,4], ocean acidification and deoxygenation [5], to increased anthropogenic perturbations from noise [6,7], overfishing and bycatch [8]. Additionally, there is an increased risk of storms and precipitation with the changing climate [9], resulting in increased freshwater events in the coastal marine environment [10,11]. This represents a conservation and management issue with respect to the species inhabiting such regions. The Gulf of Mexico, USA, is a region with significant fisheries, oil and gas industry presence and one that experiences a storm season between July and November each year [12]. In addition, this is the drainage location for the Mississippi River, which is among the highest freshwater runoffs in the world [13].

Bottlenose dolphins (*Tursiops truncatus*) are one of the best-known and well-recognized marine mammal species, found throughout temperate and tropical waters globally [14]. In the Gulf of Mexico region, multiple stocks inhabit bay, sound and estuary (BSE), coastal and offshore regions [14,15]. Distinct stocks are delineated for at least 31 BSE areas in the northern Gulf of Mexico, with three additional coastal stocks [16]. Genetic analyses support that BSE stocks are relatively discrete from one another [16,17]. BSE animals closest to shore are exposed to yearly freshwater influx from the Mississippi, Rio Grande, Mobile River and other rivers across the gulf. In addition, animals in some BSE stocks are potentially exposed to other stressors, including noise and water pollution [18–23] and a number of unusual mortality events (UME) have been documented for some BSE stocks, with possible causes (either alone or in combination) including morbillivirus, biotoxins, cold water temperatures, exposure to oil and agricultural run-off [24,25]. In addition to a challenging salinity regime, in isolated cases, individuals from some of these stocks have been displaced inshore by storm surges (i.e., moved out-of-habitat), which is cause of concern among managers [26].

The salinity in which bottlenose dolphins are typically found ranges from 20–35 parts per thousand (ppt), with a minimum of 20 ppt recommended for dolphins housed in aquaria [27]. The animals within each stock show fidelity to the estuary or embayment which they occupy, even in spite of perceived environmental challenges, which could negatively impact health [28,29]. For example, some bottlenose dolphins found in Barataria Bay in the northern Gulf of Mexico have been found to encounter salinities ranging between 1.6–32.0 ppt, spending between 1–12 consecutive days at salinities below 8 ppt [17,30,31]. However, globally, a number of studies have documented epidermal and biochemical changes associated with prolonged low salinity exposure (in both free swimming and stranded dolphins), including skin lesions, electrolyte imbalance, microbial infection and death [32–38], in addition to a disrupted prey environment affecting foraging [39]. In addition, data exist on behavioral and physiological responses in dolphins when water salinity is varied in a controlled manner or through natural events (i.e., hurricanes, floods, entrapments) [26,32,40] but die-offs have been associated with such instances [24,33,34].

Despite this knowledge base, information gaps exist regarding how different time–salinity exposures affect bottlenose dolphin health and survival. One potential method to fill this gap, while further research is undertaken, is through the use of expert elicitation. Expert elicitation is a formal, structured process in which expert knowledge of an uncertain quantity is captured in the form of a probability distribution [41]. This technique was first developed in the 1950s and 1960s [42,43], but more recently has been widely used in a range of scientific fields [44–48]. Perhaps the most high profile uses in the environmental sector have been in the assessment of risks from climate change [49] and predictions of future sea level rise [50]. In addition, expert elicitation approaches have been used previously to construct dose–response functions [51,52]. In the field of marine mammals, a number of elicitations have been conducted in recent years involving the authors and seeking to improve the methods for marine mammal issues [53,54].

The objective of this study is to combine the professional judgements of a range of experts. This method is appropriate to use where there is a relative lack of data but an urgent need for conservation or management decisions [55,56]. This approach should build upon the best available science [57]. Expert elicitation can be used to access substantive knowledge on particular topics held by experts [55], particularly to translate information obtained from multiple experts into quantitative statements that can be incorporated into a model, minimize bias in the elicited information, and ensure that uncertainty is accurately captured. Well-structured expert elicitations avoid many of the heuristics and biases that arise when experts make qualitative judgements or where such judgements are provided in an unstructured matter [57–59].

The objective of this study was to develop and parameterize a quantitative dose–response function that integrates salinity and time as the specified "dose". The intended outcome was to improve our understanding of how low salinity exposure affects dolphins, time to death (and contributing factors) and to advance the scientific foundation to support coastal management on a global scale. In addition to the results given in the main body of the paper, computer code to generate all figures and results are given in the Supplementary Materials.

2. Materials and Methods

This section has three elements. We summarize the overall elicitation approach undertaken, outline the design of this elicitation process (including the selection and preparation of experts) and describe the execution of the elicitation, the tools applied, and the statistical methods employed to generate dose–response functions.

2.1. Elicitation Approach

We employed an expert elicitation approach to develop response functions for different salinity time combinations, broadly following the Sheffield Elicitation Framework (SHELF) approach [60,61] (detailed below). This involved use of the SHELF template, carrying out introductory webinars and a formal elicitation workshop, the use of novel elicitation tools, behavioral aggregation to reach consensus via the "rational impartial observer" (RIO) approach. Crucially, this elicitation was facilitated by a trained, experienced facilitator, with support from a statistical specialist—the two authors of this study. The facilitator managed the discussion to ensure each expert engaged appropriately and the conversation was not dominated by any of the experts(s) and that the elicitation was not dominated by the common heuristics and biases that can arise, such as anchoring, availability bias and overconfidence [44].

2.2. Designing the Elicitation

The selection of experts for participation in the elicitation was based on criteria that each individual had substantial knowledge to allow provision of reliable judgements on the effects of salinity exposure on bottlenose dolphins [62] (see Supplementary Information—Table S1 for expert backgrounds). Best practice for the Sheffield framework suggests that between four and eight experts are involved in an elicitation; when elicitations are larger than this, excessive time is spent in the workshop without a corresponding increase in information being contributed [41,44]. In selecting experts, we reviewed the available literature and canvassed the research community to identify appropriate expert candidates. The final expert panel selected, comprising seven individuals, ensured a comprehensive coverage of expert judgement could be achieved across bottlenose dolphins and their ecology, and spanning the fields of epidemiology, animal physiology and veterinary science [60,63]. This coverage is a critical element of a successful elicitation as it avoids the likelihood of redundant information being presented, which may introduce bias [41,44].

Following the agreement of experts to participate, a webinar was hosted with the invited experts to further introduce the objectives of the elicitation, the formal elicitation process and discuss what should be included in the "evidence dossier", to best support their decision making Table S2.

Expert elicitation can be a mentally taxing process, even for scientists familiar with probabilities and probability distributions. This is because it is a challenge to express personal judgement as estimates with associated uncertainty. To aid and motivate the experts in advance of the workshop, and to ultimately improve the quality of the elicitation outputs [55,64], experts were asked to complete an online e-learning training course in advance of attending the in-person workshop (found at http://www.smruconsulting.com/products-tools/pcod/pcod-project-outputs/online-expert-elicitation-course, accessed on 18 November 2019). This trained the experts in subjective probabilities, distributions, making reasoned probabilistic judgements and had a series of practice exercises with bespoke feedback for the experts.

2.3. Performing the Elicitation

2.3.1. Elicitation Structure

The elicitation was conducted as an in-person workshop held at the National Ocean and Atmospheric Administration facilities, Silver Springs, USA, on the 19–21 November 2019 (see Table S2). Experts were provided with a primer on basic probability concepts including plausible limits (sometimes referred to as the 1st and 99th quantiles), median and quartiles. The facilitator used this as an opportunity to highlight and explain some of the biases and/or heuristics that can affect the quality of expert judgements—so that experts were aware of this when providing their personal judgements.

In addition to the participating experts, the elicitation was supported by scientific observers who presented foundational briefings (functioning as an "evidence dossier") on the published literature on salinity effects, unexplained mortality events (UMEs) [24], data from studies on dolphins in the US Navy Marine Mammal Program [32], a conceptual model of mechanistic pathways developed (Rowles, pers comm.) and unpublished literature/datasets available (e.g., telemetry data in different salinity regimes, relevant stranding records) to help inform judgements on quantities of interest. These observers did not provide judgements and only provided additional context when called upon.

In conjunction with the experts, the scope of the elicitation (and definitions used) was discussed and clarified. Experts were presented with a series of salinity exposure scenarios and draft questions relating to the quantity of interest. These were iteratively developed to ensure linguistic uncertainty was removed [65].

2.3.2. Low Salinity Exposure Scenarios

The elicitation was focused on three scenarios of low salinity exposure. The scenario-setting was preceded by a "scoping exercise" to focus the elicitation on plausible scenarios and pathways to impact. Realistic salinity change scenarios were developed, parameterized using salinity measurements collated from "The US Geological Survey Gulf of Mexico Dashboard" https://gom.usgs.gov/gwd (accessed on 18 November 2019). The following scenarios were considered:

- **Scenario 1A:** An extended low salinity event. For example, a bay, sound and estuary (BSE) environment (i.e., mean 15–25 ppt) is flooded with fresh or low salinity water until salinity drops (at approx. 0.5 ppt/day—i.e., salinity decreasing over 20–40 days) to below 5 ppt for an extended period. This is an environment in which animals are exposed to other significant stressors (e.g., noise, low quality prey, exposure to contaminants) and are more likely to be in a "compromised health state".
- **Scenario 1B:** As in Scenario 1A but in an environment in which there are few other stressors and animals in the population are broadly considered to be "healthy".
- **Scenario 2:** "Acute salinity change event": Bottlenose dolphins experience a change in salinity from typical salinity environment (i.e., mean 15–25 ppt) down to an atypical environment with salinity below 5 ppt for an extended period. This change in salinity occurs within 0–5 days.

Scenario 2 was designed to be applicable for events where animals are displaced by storm surges into atypical environments [26].

2.3.3. Expert Judgements

The elicitation was split into two components: the first focused on generating probability distributions of the length of exposure (d_{max}, in days) that would lead to mortality in bottlenose dolphins under a given salinity scenario, and the second focused on obtaining the parameters (μ and σ) required to determine the form (i.e., shape) of the dose–response function.

For the first component of the elicitation, a probability distribution on d_{max} was elicited separately for each scenario. Initially, the experts were asked to provide their individual subjective judgements (in the form of a probability distribution, see below) to the question: *"For the scenario defined (above), what is the length (in days) of continuous exposure to salinity below 5 ppt, that the average bottlenose dolphin in the population would need to experience to result in death (within 12 months of the start of the event)?"*. Experts discussed the potential for salinity stratification and refugia to exist in the BSE environment but agreed it would be best to elicit on the basis of continuous exposure. In addition, experts agreed to elicit for the "average" animal, to help them provide realistic judgements of what could occur in a typical population (minimizing the risk of implausible values being elicited). It was discussed with experts that this could include averaging over any factors that could cause variation in response, such as health, sex, age, etc. However, such averaging needs to take account of the expert's belief about the effect of such factors and the proportion of animals in each category (or distribution in each category for continuous variables).

Once the scenario and questions were finalized, experts provided their judgements using variable interval methods [66], first selecting their plausible range, and then bisecting this range with median and 25th and 75th quantiles. Experts used a web-based visual interface developed using the R package shiny (https://smruconsulting.shinyapps.io/EE_SingleParam, accessed on 18 November 2019) to anonymously and independently submit their judgements to facilitator for fitting to a probability distribution.

Each individual expert's judgements were fitted to probability distributions using the expert elicitation software SHELF version 3 (O'Hagan & Oakley, Sheffield, UK) [60]. This software comprised an add-on package (SHELF 1.7.0) accessed from the statistical software R 3.6.0 [67]. Within SHELF, the distribution best fitting the elicited quantiles was selected using a least-squares algorithm; the candidate distributions were normal, t, shifted gamma, lognormal, log-t, shifted scaled beta. The set of best-fitting distributions from the experts was presented back to the group and each expert was invited to provide their rationale for their judgements. To reach consensus, the group was asked to consider what a RIO may believe taking into account the individual judgements and supporting rationale. This behavioral aggregation approach helps to capture the views of multiple experts for distributions, to all experts to share and debate their opinions [41,60,61]. These rationales were discussed as a group to reach a consensus of what would be a rational impartial observer of their combined knowledge (see [41,61] for details).

The second elicitation component was to estimate the form of the relationship between survival and length of exposure to low salinity water (this form was assumed to be the same for all scenarios). To achieve this, the experts were asked to provide judgements in response to the following question: "*What form does the relationship between survival and length of exposure to low salinity take?*"

To estimate the shape of the relationship between survival and the duration of low salinity exposure, experts were provided with a tool to aid their decision making and provide their judgements. The tool was developed as a web-based Shiny application (https://lenthomas.shinyapps.io/ElicitShape2 (accessed on 2 December 2020)) and allowed elicitation of the location (μ) and shape (σ) parameters described above. Experts also rated their confidence in their elicited values on a scale of 1 (least confident) to 3 (most confident). Unlike the previous question, no uncertainty was elicited from the experts on their judgements to this question. We elicited a separate distribution from each expert, and experts agreed that a rational impartial consensus distribution would be obtained by sampling from their separate distributions.

2.4. Dose–Response Function

The resulting dose–response function is defined as follows. Let $M(d)$ be a multiplier that is applied to the baseline annual survival probability of a dolphin population as a result of d days of exposure to low salinity. $M(d)$ has a value of 1 (i.e., no effect on survival) when $d = 0$ and a value of 0 (i.e., no survival) when $d \geq d_{max}$. We define the following dose–response function

$$M(d) = 1 - \Phi_0^{100}\left(\frac{d}{d_{max}} \times 100; \mu, \sigma\right) \tag{1}$$

where $\Phi_0^{100}(x; \mu, \sigma)$ is a truncated normal cumulative distribution function with lower limit 0, upper limit 100, location parameter μ and shape parameter σ, evaluated at x (which is the percentage of the maximum days of exposure). For reference,

$$\Phi_0^{100}(x; \mu, \sigma) = \begin{cases} 0 & x < 0 \\ \frac{\Phi(x;\mu,\sigma)}{\Phi(100;\mu,\sigma) - \Phi(0;\mu,\sigma)} & 0 \leq x \leq 100 \\ 1 & x > 100 \end{cases} \tag{2}$$

where $\Phi(x; \mu, \sigma)$ is the untruncated normal cumulative distribution function with mean μ and standard deviation σ, evaluated at x. Defined this way, the dose–response function is guaranteed to be monotonic non-increasing between 0 and d_{max} days, with considerable flexibility between showing a gradual decrease (when σ is large) over that range—or part of it—through to showing a step function (when σ is small) at a particular day. This flexibility is illustrated by the range of elicited shapes shown later in the paper, and the full code to reproduce the dose–response function is available in the Supplemental Materials.

To generate dose–response relationships for each scenario, the probability distributions elicited for the scenarios above were sampled from and combined with the shape function elicited with probability proportional to expert's confidence.

3. Results

The experts were asked to identify and consider the main pathways by which low salinity exposure could impact upon the health and survival of bottlenose dolphins.

3.1. Probability Distributions

3.1.1. Extended Low Salinity Events

The first two Scenarios 1A,B, considered almost identical scenarios, with the difference between the initial health status of animals in the differing environments Figure 1. The final elicited distributions were similar for both scenarios, but with animals from an environment with few other stressors likely to be more robust to exposure. Median values were 62 days in animals exposed to other significant stressors and 77 days for animals in an environment with few other stressors. A higher concentration of mass around the median in the poorer environment distribution reflects a greater certainty of time to death, being shorter in a multiple stressor environment than in an environment with few other stressors. In both scenarios experts believed there was a small chance that short, continuous exposures (e.g., 11–12 days) could result in death of the average bottlenose dolphin, but that shorter disturbances were highly unlikely to be lethal. Experts also concluded that it was plausible that the average animal could experience much longer continuous exposures and survive, but that the other background stressors in the environment were important factors affecting the time to death (e.g., 99th percentile of up to 160 and 198 days in 1A and 1B, respectively), most clearly seen by examining the tails of the distributions Figure 1.

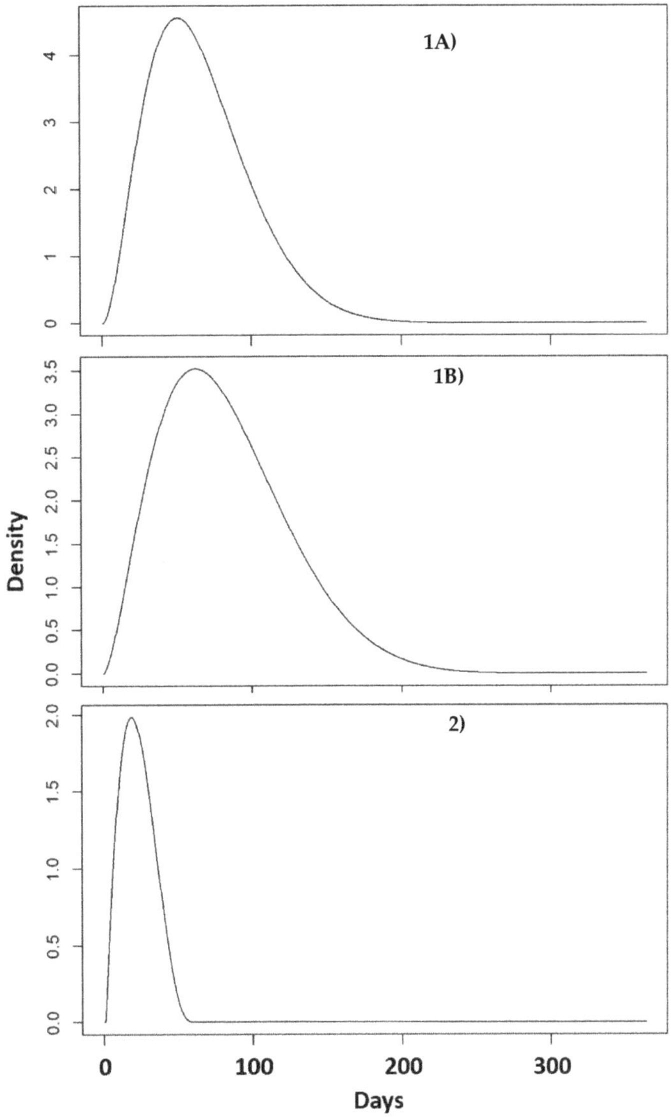

Figure 1. Elicited probability distributions for scenarios **1A** (extended low salinity event in "poor" environment, **1B** (extended low salinity event in "good" environment) and **2** (acute salinity change) showing the length in days of exposure resulting in death of the average bottlenose dolphin.

3.1.2. Acute Salinity Change Events

Experts considered that time to death would be much lower in instances where the change in salinity regime was much more acute, with a median time to death of 22 days of continuous exposure to salinity water below 5 ppt. Experts indicated that it was extremely unlikely that the average bottlenose dolphin would survive such an exposure beyond 49 days. The high concentration of mass around the median (inter-quartile range: 14–30 days) reflects the greater certainty of experts of the outcomes of acute salinity changes.

3.1.3. Expert Rationale

For all scenarios, experts considered the energetic costs associated with reduced energy intake (i.e., prey effects such as changes in density/abundance, schooling, prey type and quality) and increased energetic expenditure (e.g., dolphin buoyancy, cost of transport and reduced foraging efficiency). In addition, experts considered the age structure of populations and how they differ between healthy and unhealthy populations. The experts indicated their judgements were informed from their own research experience and the data presented and discussed in the scoping phase of the workshop.

For acute salinity change scenarios, experts noted that, in addition to the broad energetic challenges and population drivers, animals may struggle to locate prey resources. They also noted that in such instances where animals are relocated (e.g., swept by storm surges), they can be exposed to poor water quality [68,69] with limited prey availability [39], isolated from conspecifics [70] and can experience physical trauma during movement—all of which lower the duration of exposure that could be survived.

For all exposures, experts acknowledged there was the potential for animals to suffer delayed lethal effects (e.g., animals might die in subsequent years due to an initial unrecoverable exposure), but this was outside the scope of the elicitation question (which focused on effects within 12 months).

3.2. Dose–Response Function Shape

Expert judgements showed broad agreement over the shape by which continuous exposure would affect animal health (ultimately resulting in death), as shown in Figure 2. Experts agreed that animals can tolerate some exposure but that weaker animals, likely a small proportion of the population, could succumb early (e.g., young and very old animals, animals in poor health). Experts considered that the main pathway to mortality is via the skin and this likely takes some time to manifest, with the skin barrier degrading gradually as the exposure duration increases. However, once the skin barrier is compromised, a positive feedback loop exists, such that animals' condition progressively worsens, leading to increased infections, decompensation of adrenal and renal systems in addition to other chronic illnesses, and subsequent malnutrition. Experts judged that animals in the best condition at the start of low salinity exposure would die last. Figure 3 shows the resulting dose–response function from the combination of scenario-specific distributions with the generalized dose function shape, while Figures S1 and S2 show realizations drawn from these functions.

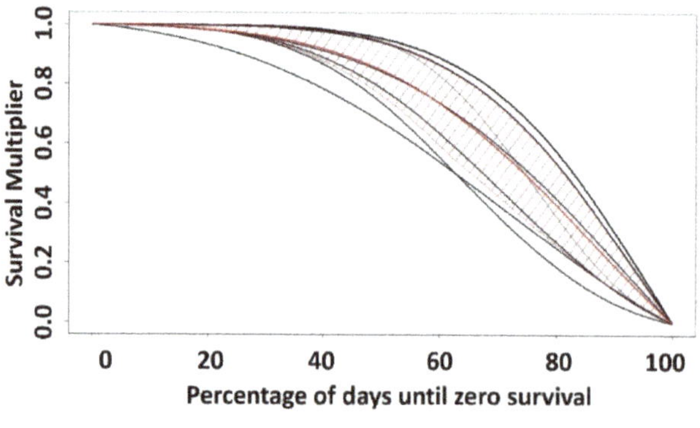

Figure 2. Elicited form of the dose–response function from each of the experts (black lines). Thickness of the line is proportional to the certainty the expert attached to their elicited shape. The red line indicates the weighted mean function, while the shaded polygon indicates approximate 50% central weighted quantiles.

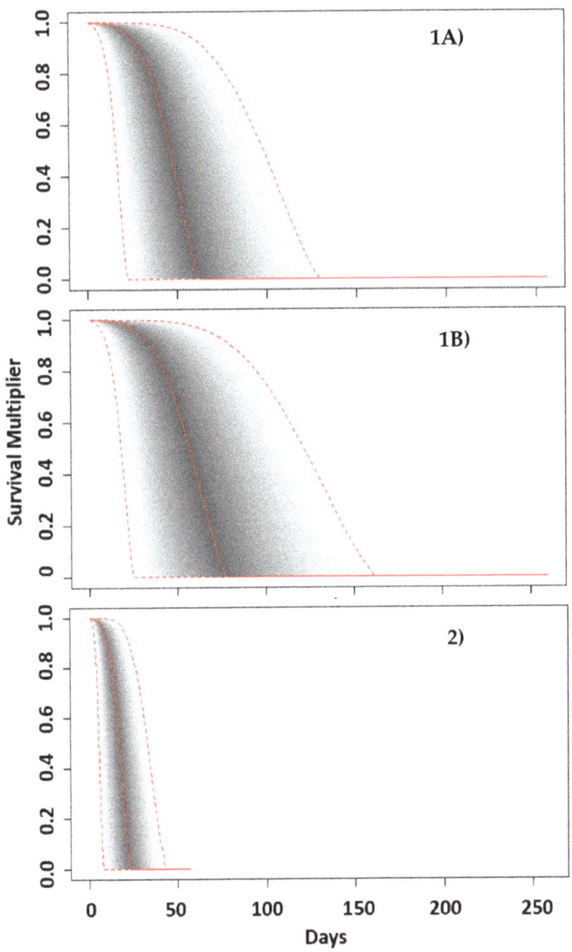

Figure 3. Dose–response functions for Scenarios **1A** (extended low salinity event in "poor" environment, **1B** (extended low salinity event in "good" environment) and **2** (acute salinity change). Figure shows the quantiles of the distribution generated by 10,000 realizations; the red solid line is the median and the dashed lines are the 5th and 95th quantiles (i.e., forming a 90% interval).

4. Discussion

This study utilizes expert knowledge to address a key management and conservation gap regarding the impacts of a changing environment on a marine mammal species. Specifically, an expert elicitation approach was employed to derive dose–response functions [51,52] and crucially characterizes the associated uncertainty and likelihood of events via probability. Such dose–response functions have applications in population modelling, quantitative risk assessments, probabilistic decision making, and for use as prior distributions in Bayesian modelling.

The dose–response functions, based on the best available science and knowledge of experts, provide an indication of how low salinity exposure may affect bottlenose dolphins. These results indicate that, in general, animals may ensure some periods of exposure to water below salinities of 5 ppt before health is impacted. This may be due to some tolerance to low salinity exposure, or perhaps more likely due to the timelines over which pathways to mortality to occur. Experts estimated these periods might be 20–30 days in the extended continuous exposure scenarios considered, but are considerably shorter (i.e., 6–8 days) for scenarios with acute changes in salinity. In all scenarios, once the survival probability began to decrease, experts indicated that they believed a relatively rapid decline in survival would occur as low salinity exposure continued. An important contextual factor in these assessments was the quality of the environment which animals inhabit and the presence of other stressors. These could negatively impact the health of individuals before low salinity exposure occurs, which has the potential to exacerbate health conditions.

The probability distributions and dose–response function generated in this study, provide the first quantitative outputs, with potential applications in management and conservation applications. For example, they could be used in an adaptive management framework to monitor and mitigate against adverse impacts. More specifically, they could be used in combination with data from the US Geological Survey (https://gom.usgs.gov/gwd, accessed on 18 November 2019) precipitation and salinity monitoring stations (or from other site-specific data collection initiatives) to predict the risk of adverse impacts on specific bottlenose dolphin stocks. This study helps provide an improved understanding of the tipping points in dolphin health and could inform when stranding monitoring might be increased and when mitigation is required. In instances where river flow into the Gulf of Mexico is regulated, the dose–response functions could be used to inform guidance thresholds for the periods over which prolonged freshwater flow is permitted. Similarly, given the predictable storm season, if dolphins are moved "out-of-habitat" [26], the acute salinity change scenario outputs could be used to inform when management action is needed.

Understanding the population level impacts of such exposures is very important. One way to achieve this is through a population simulation study, where simulated population trajectories under baseline scenarios are compared with those under scenarios, where a given proportion of the population are subjected to an altered salinity regime—changes in survival of the proportion exposed would be informed by the results given here. An example of such a study, is that which was conducted on the bottlenose dolphin population in Barataria Bay, Louisiana [20], where the impact scenarios were based on estimated changes in survival and fecundity from the Deepwater Horizon oil spill. The same population model could be used, for example, to examine possible future effects of changes in the seasonal management of the Mississippi River outflows into the bay. For this, estimates of the proportion of the population affected would be required. Similar studies could be performed on other populations, if suitable alterations to the baseline demographic parameters could be made.

We note that the results of this elicitation capture the experts' subjective judgements at the time of the workshop (and utilizing the data available at that time). The results of a duplicate elicitation with different experts or with the same group of experts at another time, could differ from those presented here. However, we do not expect them to differ significantly and they are in line with the limited available data on duration of low salinity exposures [34,35]. Astfalck et al. [61] argue in the absence of a comprehensive dataset to validate an expert elicitation, that the success of the exercise can be assessed by whether the experts are satisfied with the outcome and whether the outputs are useful. We contend that this study meets both criteria. In particular, the range of management applications described are indicative of success relating to the utility of these outputs. By utilizing the SHELF protocol [60], reviewing, collating and disseminating the available datasets and through the use of an experienced facilitator, the process of eliciting the elements to construct dose–response functions was straightforward, albeit novel in marine mammal science. The methodology presented here is transferable to the generation of any dose–response functions, provided there are adequate data to support expert judgements.

In expert elicitation it is important that there is a clear scope for experts, to aid and focus their judgements. In all scenarios, a key assumption was that animals cannot leave the low salinity areas, and therefore, the dose–response functions are for continuous exposure scenarios. Experts agreed that energetics (additional costs of inhabiting low salinity waters, whether or not exposed animals had access to suitable prey) and water quality (e.g., presence of biotoxins, pathogens, turbidity and contaminants) were considered when making their judgements. It is important to note that experts agreed not to consider the effects of temperature, stratification, or the effects of other stressors (e.g., the Deepwater Horizon oil spill [20]). In addition, as continuous exposure was the focus, any potential benefits of access to saline refugia in shallow (e.g., 1–2 m depth) and deeper BSE environments (e.g., 12–14 m depth) were not considered—though experts noted that animals might be able to access such features, but that this was poorly understood [71,72].

The study provides a means by which to temporarily fill a knowledge gap for a pressing conservation and management issue. However, a number of data gaps remain and are best addressed with additional studies of cetaceans in proximity to BSE environments. Future work is necessary to fully understand the pathways, effects, and thresholds regarding the effects of low salinity on bottlenose dolphins and other cetaceans on a global scale. Primarily, an improved understanding of the effect of aggregate exposures to repeated low salinity events, how multiple exposures within and between years are managed in the short and medium terms (e.g., freshwater pulses, short-term movements or whether there is access to potential saline refugia), and the long-term impacts of single and repeated exposures are all critical gaps. In addition, advancing the knowledge base on the physiological effects of low salinity exposure [32] is critical. Key areas include the timescales for development and/or recovery of conditions and pathways to mortality (including degeneration of the skin layer, infections through skin or gastrointestinal tract, the potential for adrenal exhaustion and renal failure). Finally, studies to improve knowledge of the bioenergetic cost of living in a low salinity environment, including how the prey base changes and how dolphin buoyancy, foraging efficiency and the costs of transport are affected. Such advancements will also help improve our understanding of the impacts of this stressor, and represent important elements to be considered along with other environmental stressors and the cumulative effects of multiple stressors (e.g., underwater noise, water temperature) on marine mammal populations [1].

Supplementary Materials: The following are available online at https://www.mdpi.com/2673-1924/2/1/11/s1, Table S1: Expert elicitation workshop participants, relevant expertise and roles; Table S2: Expert elicitation programme; Figure S1: Outputs of the dose response functions for scenarios 1A and 1B. Figure shows the first 100 realizations from dose response in Figure 3, demonstrating the range of shapes produced; Figure S2: Outputs of the dose response functions for scenario 2. Figure shows the first 100 realizations from dose response in Figure 3, demonstrating the range of shapes (note x-axis is different from Figure S1); and Extended Results (.pdf & .rmd).

Author Contributions: Conceptualization, C.B. and L.T.; methodology, C.B. and L.T.; software, C.B. and L.T.; formal analysis, C.B. and L.T.; investigation, C.B. and L.T.; writing—original draft preparation, C.B. and L.T.; writing—review and editing, C.B. and L.T.; project administration, C.B.; funding acquisition, C.B. All authors have read and agreed to the published version of the manuscript.

Funding: This research was funded by National Marine Fisheries Service for support via awards: NFFKPR00-19-01552 and NA20NMF0080281.

Informed Consent Statement: Informed consent was obtained from all subjects involved in the study.

Data Availability Statement: Data supporting the results are found in the Supplementary Materials Extended Results (.pdf & .rmd).

Acknowledgments: We would like to acknowledge the participants in the expert elicitation workshop: L.H., M.J., E.J., K.M., D.F., L.S., P.D., D.G., T.R., E.F., L.G., R.T., A.M., C.S., J.D.F., S.W., L.E., B.P., and B.L. We also would like to thank the three reviewers and editor who provided invaluable input to improve this manuscript.

Conflicts of Interest: The authors declare no conflict of interest. Funders had a role in specifying the overarching objective of the study.

References

1. National Academies of Sciences Engineering and Medicine. *Approaches to Understanding the Cumulative Effects of Stressors on Marine Mammals*; The National Academies Press: Washington, DC, USA, 2017.
2. Poloczanska, E.S.; Burrows, M.T.; Brown, C.J.; García Molinos, J.; Halpern, B.S.; Hoegh-Guldberg, O.; Kappel, C.V.; Moore, P.J.; Richardson, A.J.; Schoeman, D.S. Responses of marine organisms to climate change across oceans. *Front. Mar. Sci.* **2016**, *3*, 62. [CrossRef]
3. Frölicher, T.L.; Laufkötter, C. Emerging risks from marine heat waves. *Nat. Commun.* **2018**, *9*, 650. [CrossRef] [PubMed]
4. Harvell, C.; Montecino-Latorre, D.; Caldwell, J.; Burt, J.; Bosley, K.; Keller, A.; Heron, S.; Salomon, A.; Lee, L.; Pontier, O. Disease epidemic and a marine heat wave are associated with the continental-scale collapse of a pivotal predator (Pycnopodia helianthoides). *Sci. Adv.* **2019**, *5*, eaau7042. [CrossRef] [PubMed]
5. Bijma, J.; Pörtner, H.-O.; Yesson, C.; Rogers, A.D. Climate change and the oceans–What does the future hold? *Mar. Pollut. Bull.* **2013**, *74*, 495–505. [CrossRef]
6. King, S.L.; Schick, R.S.; Donovan, C.; Booth, C.G.; Burgman, M.; Thomas, L.; Harwood, J. An interim framework for assessing the population consequences of disturbance. *Methods Ecol. Evol.* **2015**, *6*, 1150–1158. [CrossRef]
7. Erbe, C.; Marley, S.A.; Schoeman, R.P.; Smith, J.N.; Trigg, L.E.; Embling, C.B. The Effects of Ship Noise on Marine Mammals—A Review. *Front. Mar. Sci.* **2019**, *6*. [CrossRef]
8. Davies, R.; Cripps, S.; Nickson, A.; Porter, G. Defining and estimating global marine fisheries bycatch. *Mar. Policy* **2009**, *33*, 661–672. [CrossRef]
9. Trenberth, K.E. Changes in precipitation with climate change. *Clim. Res.* **2011**, *47*, 123–138. [CrossRef]
10. Huizer, S.; Karaoulis, M.; Oude Essink, G.; Bierkens, M. Monitoring and simulation of salinity changes in response to tide and storm surges in a sandy coastal aquifer system. *Water Resour. Res.* **2017**, *53*, 6487–6509. [CrossRef]
11. Holt, T.; Seibert, S.L.; Greskowiak, J.; Freund, H.; Massmann, G. Impact of storm tides and inundation frequency on water table salinity and vegetation on a juvenile barrier island. *J. Hydrol.* **2017**, *554*, 666–679. [CrossRef]
12. Curtis, S. The Atlantic multidecadal oscillation and extreme daily precipitation over the US and Mexico during the hurricane season. *Clim. Dyn.* **2008**, *30*, 343–351. [CrossRef]
13. Nijssen, B.; O'Donnell, G.M.; Lettenmaier, D.P.; Lohmann, D.; Wood, E.F. Predicting the discharge of global rivers. *J. Clim.* **2001**, *14*, 3307–3323. [CrossRef]
14. Vollmer, N.L.; Rosel, P.E. A review of common bottlenose dolphins (Tursiops truncatus truncatus) in the northern Gulf of Mexico: Population biology, potential threats, and management. *Southeast Nat.* **2013**, *12*, 1–43.
15. Hayes, S.; Josephson, E.; Maze-Foley, K.; Rosel, P. US Atlantic and Gulf of Mexico marine mammal stock assessments–2018. *NOAA Tech. Memo. NMFS-NE* **2019**, *258*. [CrossRef]
16. NOAA Fisheries. *Common Bottlenose Dolphin (Tursiops Truncatus Truncatus): Northern Gulf of Mexico Bay, Sound, and Estuary Stocks*; NOAA Southeast Fisheries Science Center: Miama, FL, USA, 2019; p. 21.
17. Rosel, P.; Wilcox, L.; Sinclair, C.; Speakman, T.; Tumlin, M.; Litz, J.; Zolman, E. Genetic assignment to stock of stranded common bottlenose dolphins in southeastern Louisiana after the Deepwater Horizon oil spill. *Endanger. Species Res.* **2017**, *33*, 221–234. [CrossRef]
18. Balmer, B.C.; Schwacke, L.H.; Wells, R.S.; George, R.C.; Hoguet, J.; Kucklick, J.R.; Lane, S.M.; Martinez, A.; McLellan, W.A.; Rosel, P.E.; et al. Relationship between persistent organic pollutants (POPs) and ranging patterns in common bottlenose dolphins (*Tursiops truncatus*) from coastal Georgia, USA. *Sci. Total Environ.* **2011**, *409*, 2094–2101. [CrossRef]
19. Rowles, T.K.; Schwacke, L.S.; Wells, R.S.; Saliki, J.T.; Hansen, L.; Hohn, A.; Townsend, F.; Sayre, R.A.; Hall, A.J. Evidence of susceptibility to morbillivirus infection in cetaceans from the United States. *Mar. Mammal Sci.* **2011**, *27*, 1–19. [CrossRef]
20. Schwacke, L.H.; Thomas, L.; Wells, R.S.; McFee, W.E.; Hohn, A.A.; Mullin, K.D.; Zolman, E.S.; Quigley, B.M.; Rowles, T.K.; Schwacke, J.H. Quantifying injury to common bottlenose dolphins from the Deepwater Horizon oil spill using an age-, sexand class-structured population model. *Endang. Species Res.* **2017**, *33*, 265–279. [CrossRef]
21. Takeshita, R.; Sullivan, L.; Smith, C.; Collier, T.; Hall, A.; Brosnan, T.; Rowles, T.; Schwacke, L. The Deepwater Horizon oil spill marine mammal injury assessment. *Endang. Species Res.* **2017**, *33*, 95–106. [CrossRef]
22. Venn-Watson, S.K.; Townsend, F.I.; Daniels, R.L.; Sweeney, J.C.; Mcbain, J.W.; Klatsky, L.J.; Hicks, C.L.; Staggs, L.A.; Rowles, T.K.; Schwacke, L.H.; et al. Hypocitraturia in Common Bottlenose Dolphins (*Tursiops truncatus*): Assessing a Potential Risk Factor for Urate Nephrolithiasis. *Comp. Med.* **2010**, *60*, 149–153.
23. Mchugh, K.A.; Allen, J.B.; Barleycorn, A.A.; Wells, R.S. Severe Karenia brevis red tides influence juvenile bottlenose dolphin (*Tursiops truncatus*) behavior in Sarasota Bay, Florida. *Mar. Mammal Sci.* **2011**, *27*, 622–643. [CrossRef]
24. Litz, J.A.; Baran, M.A.; Bowen-Stevens, S.R.; Carmichael, R.H.; Colegrove, K.M.; Garrison, L.P.; Fire, S.E.; Fougeres, E.M.; Hardy, R.; Holmes, S. Review of historical unusual mortality events (UMEs) in the Gulf of Mexico (1990–2009): Providing context for the multi-year northern Gulf of Mexico cetacean UME declared in 2010. *Dis. Aquat. Org.* **2014**, *112*, 161–175. [CrossRef]
25. Carmichael, R.H.; Graham, W.M.; Aven, A.; Worthy, G.; Howden, S. Were Multiple Stressors a 'Perfect Storm' for Northern Gulf of Mexico Bottlenose Dolphins (*Tursiops truncatus*) in 2011? *PLoS ONE* **2012**, *7*. [CrossRef]
26. Rosel, P.E.; Watts, H. Hurricane impacts on bottlenose dolphins in the northern Gulf of Mexico. *Gulf Mex. Sci.* **2007**, *25*, 88. [CrossRef]
27. Andersen, S. Treatment of water in dolphinaria. *Aquat. Mamm* **1973**, *1*, 1–18.

28. Wells, R.S.; Rhinehart, H.L.; Hansen, L.J.; Sweeney, J.C.; Townsend, F.I.; Stone, R.; Casper, D.R.; Scott, M.D.; Hohn, A.A.; Rowles, T.K. Bottlenose dolphins as marine ecosystem sentinels: Developing a health monitoring system. *EcoHealth* **2004**, *1*, 246–254. [CrossRef]
29. Hart, L.B.; Rotstein, D.S.; Wells, R.S.; Allen, J.; Barleycorn, A.; Balmer, B.C.; Lane, S.M.; Speakman, T.; Zolman, E.S.; Stolen, M. Skin lesions on common bottlenose dolphins (Tursiops truncatus) from three sites in the Northwest Atlantic, USA. *PLoS ONE* **2012**, *7*, e33081. [CrossRef]
30. McDonald, T.L.; Hornsby, F.E.; Speakman, T.R.; Zolman, E.S.; Mullin, K.D.; Sinclair, C.; Rosel, P.E.; Thomas, L.; Schwacke, L.H. Survival, density, and abundance of common bottlenose dolphins in Barataria Bay (USA) following the Deepwater Horizon oil spill. *Endanger. Species Res.* **2017**, *33*, 193–209. [CrossRef]
31. Hornsby, F.E.; McDonald, T.L.; Balmer, B.C.; Speakman, T.R.; Mullin, K.D.; Rosel, P.E.; Wells, R.S.; Telander, A.C.; Marcy, P.W.; Klaphake, K.C. Using salinity to identify common bottlenose dolphin habitat in Barataria Bay, Louisiana, USA. *Endanger. Species Res.* **2017**, *33*, 181–192. [CrossRef]
32. McClain, A.M.; Daniels, R.; Gomez, F.M.; Ridgway, S.H.; Takeshita, R.; Jensen, E.D.; Smith, C.R. Physiological Effects of Low Salinity Exposure on Bottlenose Dolphins (Tursiops truncatus). *J. Zool. Bot. Gard.* **2020**, *1*, 61–75. [CrossRef]
33. Colbert, A.A.; Scott, G.; Fulton, M.; Wirth, E.; Daugomah, J.; Key, P.; Strozier, E.; Galloway, S. *Investigation of Unusual Mortalities of Bottlenose Dolphins Along the Mid-Texas Coastal Bay Ecosystem During 1992*; US Department of Commerce: Washington, DC, USA, 1999.
34. Mullin, K.; Barry, K.P.; Sinclair, C.; Litz, J.A.; Maze-Foley, K.; Fougères, E.M.; Ewing, R.; Gorgone, A.M.; Adams, J.; Tumlin, M. *Common Bottlenose Dolphins (Tursiops Truncatus) in Lake Pontchartrain, Louisiana, 2007 to Mid-2014*; NOAA Southeast Fisheries Science Center: Miama, FL, USA, 2015.
35. Barry, K.P.; Gorgone, A.M.; Mase, B. Lake Pontchartrain, Louisiana Bottlenose Dolphin Survey Summary 28 April 2008–10 May 2008. In *Southeast Fisheries Science Center, National Marine Fisheries Service, NOAA Protected Resources and Biodiversity Division, PRBD Contribution: PRBD-08/09-01*; NOAA Southeast Fisheries Science Center: Miama, FL, USA, 2008.
36. Holyoake, C.; Finn, H.; Stephens, N.; Duignan, P.; Salgado, C.; Smith, H.; Bejder, L.; Linke, T.; Daniel, C.; Lo, H. *Technical Report on the Bottlenose Dolphin (Tursiops Aduncus) Unusual Mortality Event within the Swan Canning Riverpark, June–October 2009*; Murdoch University: Perth, Australia, 2010.
37. Rowe, L.E.; Currey, R.J.; Dawson, S.M.; Johnson, D. Assessment of epidermal condition and calf size of Fiordland bottlenose dolphin Tursiops truncatus populations using dorsal fin photographs and photogrammetry. *Endanger. Species Res.* **2010**, *11*, 83–89. [CrossRef]
38. Duignan, P.J.; Stephens, N.S.; Robb, K. Fresh water skin disease in dolphins: A case definition based on pathology and environmental factors in Australia. *Sci. Rep.* **2020**, *10*, 1–17. [CrossRef] [PubMed]
39. Smith, C.E.; Hurley, B.J.; Toms, C.N.; Mackey, A.D.; Solangi, M.; Kuczaj, S.A. Hurricane impacts on the foraging patterns of bottlenose dolphins *Tursiops truncatus* in Mississippi Sound. *Mar. Ecol. Prog. Ser.* **2013**, *487*, 231–244. [CrossRef]
40. Fandel, A.D.; Garrod, A.; Hoover, A.; Wingfield, J.; Lyubchich, V.; Secor, D.; Hodge, K.; Rice, A.; Bailey, H. Effects of intense storm events on dolphin occurrence and foraging behavior. *Sci. Rep.* **2020**, *10*, 1–9. [CrossRef]
41. O'Hagan, A. Expert knowledge elicitation: Subjective but scientific. *Am. Stat.* **2019**, *73*, 69–81. [CrossRef]
42. Brown, B.B. *Delphi Process: A Methodology Used for the Elicitation of Opinions of Experts*; Rand Corp.: Santa Monica, CA, USA, 1968.
43. O'Hagan, A.; Buck, C.E.; Daneshkhah, A.; Eiser, J.R.; Garthwaite, P.H.; Jenkinson, D.J.; Oakley, J.E.; Rakow, T. *Uncertain Judgements: Eliciting Experts' Probabilities*; John Wiley & Sons: Hoboken, NJ, USA, 2006.
44. European Food Safety Authority. *Guidance on Expert Knowledge Elicitation in Food and Feed Safety Risk Assessment*; European Food Safety Authority (EFSA): Parma, Italy, 2014.
45. Knol, A.B.; Slottje, P.; van der Sluijs, J.P.; Lebret, E. The use of expert elicitation in environmental health impact assessment: A seven step procedure. *Environ. Health* **2010**, *9*, 19. [CrossRef]
46. MacMillan, D.C.; Marshall, K. The Delphi process–an expert-based approach to ecological modelling in data-poor environments. *Anim. Conserv.* **2006**, *9*, 11–19. [CrossRef]
47. Sivle, L.D.; Kvadsheim, P.H.; Curé, C.; Isojunno, S.; Wensveen, P.J.; Lam, F.-P.A.; Visser, F.; Kleivane, L.; Tyack, P.L.; Harris, C.M. Severity of expert-identified behavioural responses of humpback whale, minke whale, and northern bottlenose whale to naval sonar. *Aquat. Mamm.* **2015**, *41*, 469. [CrossRef]
48. Aspinall, W. A route to more tractable expert advice. *Nature* **2010**, *463*, 294. [CrossRef] [PubMed]
49. Lenton, T.M.; Held, H.; Kriegler, E.; Hall, J.W.; Lucht, W.; Rahmstorf, S.; Schellnhuber, H.J. Tipping elements in the Earth's climate system. *Proc. Natl. Acad. Sci. USA* **2008**, *105*, 1786–1793. [CrossRef]
50. Bamber, J.L.; Aspinall, W. An expert judgement assessment of future sea level rise from the ice sheets. *Nat. Clim. Chang.* **2013**, *3*, 424. [CrossRef]
51. Albert, I.; Donnet, S.; Guihenneuc-Jouyaux, C.; Low-Choy, S.; Mengersen, K.; Rousseau, J. Combining expert opinions in prior elicitation. *Bayesian Anal.* **2012**, *7*, 503–532. [CrossRef]
52. Boobis, A.; Flari, V.; Gosling, J.P.; Hart, A.; Craig, P.; Rushton, L.; Idahosa-Taylor, E. Interpretation of the margin of exposure for genotoxic carcinogens–Elicitation of expert knowledge about the form of the dose response curve at human relevant exposures. *Food Chem. Toxicol.* **2013**, *57*, 106–118. [CrossRef]

53. Booth, C.; Burgman, M.; Donovan, C.; Harwood, J.; Thomas, L.; Schick, R.; Wood, J. *PCoD Lite-Using an Interim PCoD Protocol to Assess the Effects of Disturbance Associated with US Navy Exercises on Marine Mammal Populations*; DTIC Document: St Andrews, UK, 2014.
54. Tollit, D.; Harwood, J.; Booth, C.; Thomas, L.; New, L.F.; Wood, J. *Cook Inlet Beluga Whale PCoD Expert Elicitation Workshop Report. Prepared by SMRU Consulting North America for NOAA Fisheries*; SMRU Consulting: Friday Harbor, WA, USA, 2016.
55. Martin, T.G.; Burgman, M.A.; Fidler, F.; Kuhnert, P.M.; Low-Choy, S.; McBride, M.; Mengersen, K. Eliciting expert knowledge in conservation science. *Conserv. Biol.* **2012**, *26*, 29–38. [CrossRef]
56. Runge, M.C.; Converse, S.J.; Lyons, J.E. Which uncertainty? Using expert elicitation and expected value of information to design an adaptive program. *Biol. Conserv.* **2011**, *144*, 1214–1223. [CrossRef]
57. Morgan, M.G. Use (and abuse) of expert elicitation in support of decision making for public policy. *Proc. Natl. Acad. Sci. USA* **2014**, *111*, 7176–7184. [CrossRef] [PubMed]
58. Kynn, M. The 'heuristics and biases' bias in expert elicitation. *J. R. Stat. Soc. Ser. Stat. Soc.* **2008**, *171*, 239–264. [CrossRef]
59. Kahneman, D. *Thinking, Fast and Slow*; Macmillan: London, UK, 2011.
60. Gosling, J.P. SHELF: The Sheffield elicitation framework. In *Elicitation*; Springer: Berlin/Heidelberg, Germany, 2018; pp. 61–93.
61. Astfalck, L.; Cripps, E.; Gosling, J.; Hodkiewicz, M.; Milne, I. Expert elicitation of directional metocean parameters. *Ocean. Eng.* **2018**, *161*, 268–276. [CrossRef]
62. Pasanisi, A.; Keller, M.; Parent, E. Estimation of a quantity of interest in uncertainty analysis: Some help from Bayesian decision theory. *Reliab. Eng. Syst. Saf.* **2012**, *100*, 93–101. [CrossRef]
63. Hart, A.; Gosling, J.P.; Quigley, J.; Revie, M.; Thulke, H.H.; Brock, T.; O'Hagan, T.; Charlton, C. Training courses on "Steering an Expert Knowledge Elicitation" and "Use of the Expert Knowledge Elicitation Guidance in Risk Assessments for EFSA Management" and "Conduct of the Sheffield protocol for an Expert Knowledge Elicitation". *EFSA Supporting Publ.* **2018**, *15*, 1450E. [CrossRef]
64. Renooij, S. Probability elicitation for belief networks: Issues to consider. *Knowl. Eng. Rev.* **2001**, *16*, 255. [CrossRef]
65. Carey, J.M.; Burgman, M.A. Linguistic uncertainty in qualitative risk analysis and how to minimize it. *Ann. N. Y. Acad. Sci.* **2008**, *1128*, 13–17. [CrossRef]
66. Hora, S.C. Acquisition of expert judgment: Examples from risk assessment. *J. Energy Eng.* **1992**, *118*, 136–148. [CrossRef]
67. *R Core Team R: A Language and Environment for Statistical Computing*; R Foundation for Statistical Computing: Vienna, Austria, 2016.
68. Fazioli, K.L.; Hofmann, S.; Wells, R.S. Use of Gulf of Mexico coastal waters by distinct assemblages of bottlenose dolphins (Tursiops truncatus). *Aquat. Mamm.* **2006**, *32*, 212. [CrossRef]
69. Fury, C.A.; Harrison, P.L. Impact of flood events on dolphin occupancy patterns. *Mar. Mammal Sci.* **2011**, *27*, E185–E205. [CrossRef]
70. Hoffland, T.; Yeater, D.B.; Kuczaj II, S.A.; Solangi, M. Importance of social relationships in a group of Bottlenose dolphins (Tursiops truncatus) during a natural disaster. *Aquat. Mamm.* **2017**, *43*, 391. [CrossRef]
71. Hickey, B.; MacCready, P.; Elliott, E.; Kachel, N. Dense saline plumes in Exuma sound, Bahamas. *J. Geophys. Res. Ocean.* **2000**, *105*, 11471–11488. [CrossRef]
72. Manzello, D.P.; Enochs, I.C.; Melo, N.; Gledhill, D.K.; Johns, E.M. Ocean acidification refugia of the Florida Reef Tract. *PLoS ONE* **2012**, *7*, e41715.

MDPI
St. Alban-Anlage 66
4052 Basel
Switzerland
Tel. +41 61 683 77 34
Fax +41 61 302 89 18
www.mdpi.com

Oceans Editorial Office
E-mail: oceans@mdpi.com
www.mdpi.com/journal/oceans

www.ingramcontent.com/pod-product-compliance
Lightning Source LLC
Chambersburg PA
CBHW062105220926
51850CB00010B/2899